Revise A2 Geography

Author
Peter Goddard

Contents

Specification lists

AQA A Geography

UNIT	SPECIFICATION TOPIC	CHAPTER REFERENCE	STUDIED IN CLASS	REVISED	PRACTICE QUESTIONS
Unit 4 (U4) *Challenge and change in the natural environment*	***Two from:***				
	Coasts – processes and problems	*AS 2.1, 2.2 A2 Chapter 7*			
	Geomorphological processes and hazards	*AS 4.1, 4.2 A2 Chapter 8*			
	Cold environments and human activity	*AS 1.1, 1.2*			
Unit 5 (U5) *Challenge and change in the natural environment*	***Two from:***				
	Population pressure and resource management	*AS 7.1–7.4 A2 Chapter 8*			
	Managing cities: challenges and issues	*AS 6.1–6.3 A2 Chapter 8*			
	Recreation and tourism	*AS 6.1–6.4*			
Unit 6 (U6) *or* **Unit 7 (U7)**	*Fieldwork investigation*	*See below*			
	Written fieldwork investigation	*See below*			

Examination analysis

Module 4	*Two questions from three physical structured questions, plus one essay (synoptic) from a choice of three. Content drawn from Unit 4.*	*1 hr 30 min test*	*15%*
Module 5	*Two questions from three human structured questions, plus one essay (synoptic) from a choice of three. Content drawn from Unit 5.*	*1 hr 30 min test*	
Either			
Module 6	*Board assessed individual study. 4000 words from any aspect/area of the specification.*		*20%*
or			
Module 7	*Written fieldwork paper, under exam conditions, using pre-released material. Topics published in advance.*	*2 hr test*	*20%*

AQA B Geography

UNIT	SPECIFICATION TOPIC	CHAPTER REFERENCE	STUDIED IN CLASS	REVISED	PRACTICE QUESTIONS
Unit 4 (U4) Global change	**Physical geography**				
	Atmospheric change	AS 3.1, 3.2 A2 Chapter 8			
	Soils	AS 5.1 A2 Chapter 8			
	Tectonics	AS 5.1 A2 Chapter 8			
	People and the environment				
	Human Geography				
	Global economy	AS 5.1–5.5			
	Migration	AS 7.1–7.4 A2 4.1–4.3			
	The EU	A2 Chapter 8			
	Separatism				
Unit 5 (U5) The synoptic module	Based on 30+ specified topics	A2 Chapter 8			
Unit 6 (U6) or Unit 7 (U7)	Based on a knowledge of how to undertake a geographical inquiry	See below			
	An investigative report based on primary data collection	See below			

Examination analysis

Module 4	One of two structured questions and one of two essay questions based on A2 Unit 4.	1 hr 30 min test	15%
Module 5	Based on an advance released information booklet. All structured questions are answered. They will be of a decision-making format.	1 hr 30 min test	20%
Either			
Module 6	Two from two structured questions	1 hr 30 min test	15%
or			
Module 7	Investigation of 3500 to 4000 words. Centre assessed.		15%

OCR A Geography

UNIT	SPECIFICATION TOPIC	CHAPTER REFERENCE	STUDIED IN CLASS	REVISED	PRACTICE QUESTIONS
Unit 4 (U4) *Options in physical geography*	***One from Physical list:***				
	Coasts	*AS 2.1, 2.2*			
	Rivers	*AS 1.1–1.3*			
	Glaciation and periglaciation	*A2 1*			
	Hot and semi arid environments	*A2 3.1–3.4*			
	Applied climatology	*AS 3.1, 3.2* *A2 Chapter 8*			
	One from Human list:				
	Agriculture and food	*AS 7.1–7.4*			
	Manufacturing industry	*AS 8.1–8.3*			
	Service industry	*AS 8.1–8.3*			
	Tourism and recreation	*AS 6.1–6.4*			
Unit 5 (U5) *People and environment options*	***Two from:***				
	Geographical aspects of the European Union	*A2 Chapter 8*			
	Managing the urban environment	*AS 6.1–6.3*			
	Managing the rural environment	*AS 6.1–6.3* *A2 7.1–7.4, Chapter 8*			
	Hazardous environments	*AS 4.1, 4.2* *A2 Chapter 8*			
Unit 6 (U6) *Personal investigative study* ***or*** **Unit 7 (U7)** *Investigative skills paper*		*See below*			

Examination analysis

Module 4 **Written examination**. *For each of the physical and human options three essays are set. Students produce essays from each of the two options, one from the physical list and one from the human list.* — *1 hr 30 min test* — *15%*

Module 5 **Written examination (synoptic)**. *In this unit there are three essay questions for each option. Candidates produce one essay for each of any two options.* — *1 hr 30 min test* — *20%*

Either

Module 6 **Coursework** – *personal investigative study. 2500 words that relates to the specification.* — *15%*

or

Module 7 **Written exam and coursework**. *1000 word report/inquiry investigating an issue or problem based on primary data that is analysed and concluded. In the examination students answer one question related to the inquiry and one question that relates to the 'essential' elements of the inquiry.* — *1 hr 30 min test* — *15%*

OCR B Geography

UNIT	SPECIFICATION TOPIC	CHAPTER REFERENCE	STUDIED IN CLASS	REVISED	PRACTICE QUESTIONS
Unit 4 (U4) A geographical investigation	See below				
Unit 5 (U5) Issues in the environment	**Physical options (choose two):**				
	Natural hazards and human responses	AS 4.1 A2 Chapter 8			
	Climate and society	AS 3.1, 3.2 A2 Chapter 8			
	Cold environment and human responses	A2 1.1, 1.2			
	Tropical environment and people	A2 2.1, 2.2			
	Human options (choose two):				
	Food supply	A2 7.1–7.4, Chapter 8			
	Changing urban places	AS 6.1–6.3			
	Leisure and tourism	A2 6.1–6.4			
	Globalisation of economic activity	AS 8.1–8.3 A2 5.1–5.5, Chapter 8			
Unit 6 (U6) Issues in sustainable development	Global energy resources	A2 Chapter 8			
	Fresh water supplies				
	Oceans as a threatened resource				
	Soil degradation				
	Forest utilisation				
	Air quality and health				
	Landscape				
	Transport				
	Waste disposal and minimalisation				

Examination analysis

Module 4 **Coursework** – this can be based on fieldwork or the use of ICT and the production of a report. 15%

Module 5 **Written examination**. Candidates select one option from a list of physical options and one from a list of human options. Each option will have two essay type questions. The first essay question will be based on stimulus material, while there will be some choice with the second question. One essay is answered for each of the option lists. 2 hr test 15%

Module 6 **Written examination (synoptic)**. This is based on pre-released data on a selected issue and contains a variable number of compulsory structured questions and tasks. Content is principally based on Unit 6 content. 1 hr 30 min test 20%

Edexcel A Geography

UNIT	SPECIFICATION TOPIC	CHAPTER REFERENCE	STUDIED IN CLASS	REVISED	PRACTICE QUESTIONS
Unit 4 (U4) *Physical systems, processes and patterns*	***Could select two from:***				
	Atmospheric systems	*AS 3.1, 3.2 A2 Chapter 8*			
	Glacial systems	*A2 1.1, 1.2*			
	Ecosystems and soil	*AS 5.1, 5.2 A2 Chapter 8*			
Unit 5 (U5) *Human systems, processes and patterns*	***Could select two from:***				
	Economic systems	*AS 6.1–6.3 A2 Chapter 8*			
	Rural urban development	*AS 6.1–6.3 A2 Chapter 8*			
	Development processes	*A2 5.1–5.5*			
Unit 6 (U6) *The Synoptic unit: people and their environment*	***Themes include:***	*A2 Chapter 8*			
	Physical environment influences human activity				
	Human activity modifies the physical environment				
	Physical and human resources may be exploited, managed and protected				
	Communities and their governance influence geographical inter-relationships at a range of scales				

Examination analysis

Module 4 ***Written examination.*** *Six semi-structured essay questions are set. Stimulus resources will be offered. Two questions are set on each of the three areas of study. Two questions from different sections to be answered.* — *1 hr 30 min test, 15%*

Module 5 ***Written examination.*** *Six semi-structured essay questions are set. Stimulus resources will be offered. Two questions are set on each of the three areas of study. Two questions from different sections to be answered. Candidates may be asked to use maps and diagrams, and draw on their personal investigation in their answers.* — *1 hr 30 min test, 15%*

Module 6 ***Written examination.*** *Two sections. The first is a compulsory section, it is sub-divided into sub-sections; candidates are expected to draw on their knowledge, understanding and skills in geography, in an unfamiliar situation. In the second section four synoptic essays are set, one has to be answered.* — *2 hr test, 20%*

Edexcel B Geography

UNIT	SPECIFICATION TOPIC	CHAPTER REFERENCE	STUDIED IN CLASS	REVISED	PRACTICE QUESTIONS
Unit 4 (U4) Global challenge	***The natural environment***				
	Atmospheric processes	*AS 3.1, 3.2 A2 Chapter 8*			
	Ecosystems under pressure	*AS 5.1, 5.2 A2 2.1, 2.2, Chapter 8*			
	Population and the economy				
	Global population and migration	*AS 7.1–7.4 A2 4.1–4.3, Chapter 8*			
	Globalisation and the global economy	*AS 8.1–8.3 A2 Chapter 8*			
Unit 5 (U5) Researching global futures	***Managing the natural environment** One from:*				
	The environment and resources	*A2 Chapter 8*			
	Living with the hazardous environment	*AS 4.1, 4.2 A2 Chapter 8*			
	The pollution of the natural environment	*A2 Chapter 8*			
	***Challenge for the human environment** One from:*				
	Development and disparity	*A2 5.1–5.5*			
	Feeding the world's people	*A2 7.1–7.4, Chapter 8*			
	Health and welfare	*A2 Chapter 5*			
	The geography of sport and leisure	*A2 6.1–6.4*			
Unit 6 (U6) Synoptic unit: issues analysis	***Unit 1** – Changing landforms and management*	*AS and A2 see below*			
	***Unit 2** – Changing human environments*				
	***Unit 3** – Global challenge*				

Examination analysis

Module 4	***Written examination.** Students choose three structured, data and stimulus essay style questions, one from the natural environment and one from the human section. A third is chosen from a cross-modular section. A choice of questions is set in each section.*	*2 hr test 15%*
Module 5	***Written examination and coursework report.** Managing the natural environment is assessed with one formal research-based essay from two, both questions will be based on a generalisation that would have been pre-released. 1500 word test. The challenge of the human environment is assessed with a coursework report. Students choose from a published list. The 1500 word unit is assessed externally.*	*80 min 15%*
Module 6	***Written examination.** Assessed by an issues analysis exercise. Linked tasks test critical understanding the knowledge, understanding and skills drawn from Units 1, 2 and 4. Students are assessed on their ability to synthesise, make decisions and problem-solve. Resources for the exam are pre-released.*	*2 hr test 20%*

WJEC Geography

UNIT	SPECIFICATION TOPIC	CHAPTER REFERENCE	STUDIED IN CLASS	REVISED	PRACTICE QUESTIONS
Unit 4 (U4) *Geographical processes and their management*	***Two from four options***				
	Landforms: processes and management	*Throughout AS and A2*			
	Climate hazards: causes and management	*AS 3.1, 3.2 A2 Chapter 8*			
	Inequalities in development	*AS 8.1–8.3 A2 5.1–5.5, Chapter 8*			
	The changing geography of economic development	*AS 8.1–8.3 A2 5.1–5.5, Chapter 8*			
Unit 5 (U5) *Sustainable development*	***Sustainability and:***				
	Food supply	*A2 7.1–7.4, Chapter 8*			
	Water supply	*AS 1.1–1.3 A2 Chapter 8*			
	The natural environment	*Throughout AS and A2*			
	The urban environment	*Throughout AS and A2*			
Unit 6a (U6a) *Personal investigation* **or** **Unit 6b (U6b)** *Geographical assignment*	*See below*				

Examination analysis

Module 4	***Written examination**. Two essays, one from a choice within each of the options.*	*1 hr 30 min test*	*15%*
Module 5	***Written examination and decision-making exercise.*** *Includes 30 minutes reading time.* ***Section A:*** *Essay (synoptic in nature).* ***Section B:*** *Decision making exercise.*	*3 hr test*	*20%*
Either ***Module 6 (a)***	***Individual investigation*** *– internally assessed.*		*15%*
or ***Module 6 (b)***	***Geographical investigation*** *– externally assessed.*		*15%*

NICCEA Geography

UNIT	SPECIFICATION TOPIC	CHAPTER REFERENCE	STUDIED IN CLASS	REVISED	PRACTICE QUESTIONS
Unit 4 (U4) *Physical processes and human interactions*	***Choose two from:***				
	Managing fluvial and coastal environments	*AS 1.1–2.2 A2 Chapter 8*			
	The nature and vulnerability of tropical ecosystems	*A2 2.1, 2.2*			
	Pollution and its management	*A2 Chapter 8*			
Unit 5 (U5) *Processes and issues in human geography*	***Choose two from:***				
	Impact of population change	*AS 7.1–7.4 A2 4.1–4.3, Chapter 8*			
	Issues in ethnic diversity	—			
	Planning for sustainable development	*Throughout AS and A2*			
	The changing nature of economic activity	*AS 8.1–8.3 A2 Chapter 8*			
Unit 6 (U6) *Skills and decision making in geography*		*See below*			

Examination analysis

Module 4	***Written examination and structured questions.*** *Students do two questions, one from each of their two chosen sections. Each question will have an extended end section (5% synoptic assessment).*	*1 hr 30 min test 15%*
Module 5	***Written examination and structured questions.*** *Students answer two questions, one from each of their two chosen sections. Each question will have an extended end section.*	*1 hr 30 min test 15%*
Module 6	***Written skills and decision making examination.*** *In Section A candidates answer a compound skills based question – drawing on skills and information that is provided and based on experience of fieldwork. In Section B candidates undertake a decision-making exercise. Resources are offered (15% synoptic assessment).*	*2 hr test 20%*

AS/A2 Level Geography courses

AS and A2

Since September 2000 all A Level qualifications comprised three units of AS assessment and three units of A2 assessment. This offers Geography students the opportunity to complete a freestanding AS course or to complete their geographical education, and to develop ideas, themes and concepts further into a full A Level course via the much more demanding and challenging A2 course.

How will you be tested?

Assessment units

For AS Geography, you will be tested by three assessment units. For the full A Level in Geography, you will take a further three A2 units.

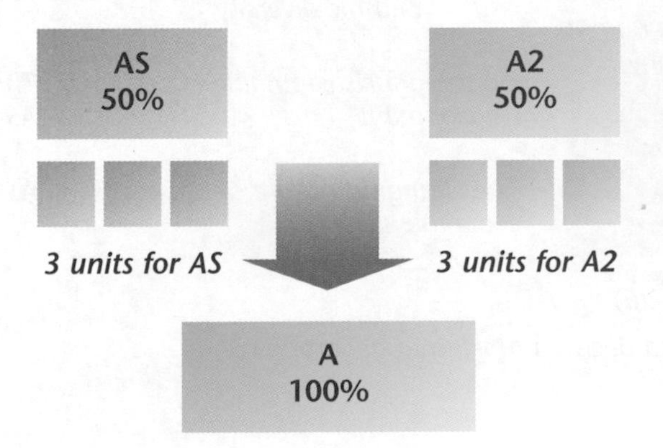

Each unit can normally be taken in either January or June. Alternatively, you can study the whole course before taking any of the unit tests. There is a lot of flexibility about when exams can be taken and the diagram below shows just some of the ways that the assessment units may be taken for AS and A2 Level Geography.

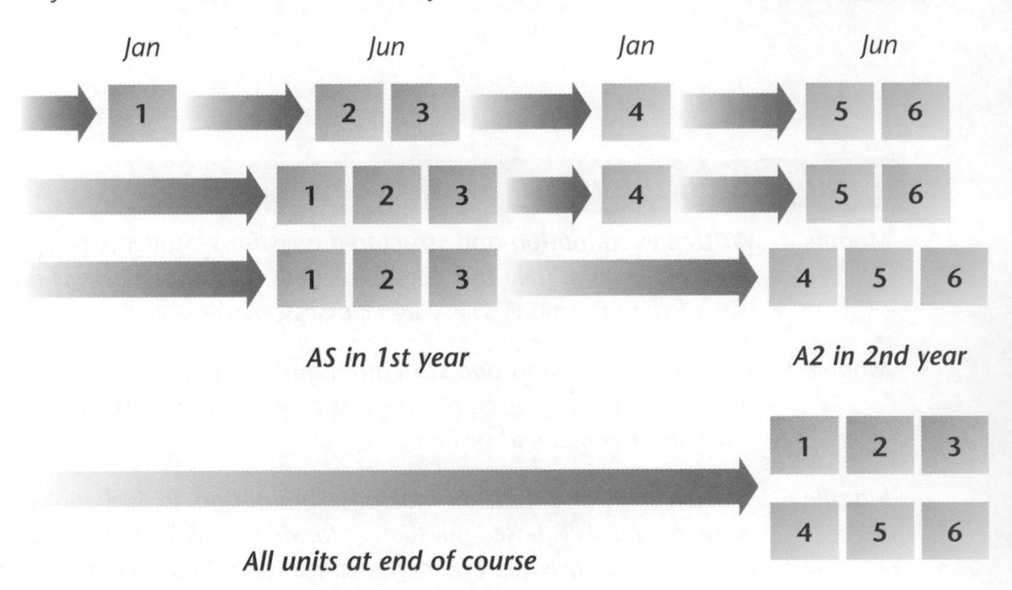

If you are disappointed with a module result, you can resit each module once. You will need to be very careful about when you take up a resit opportunity because you will have only one chance to improve your mark. The higher mark counts.

A2 and synoptic assessment

Having studied AS Geography you are now studying Geography to A2 Level. For this assessment you will need to take three further units of Geography. Similar assessment arrangements to AS apply except some units, those that draw together different parts of the course in a synoptic assessment, have to be assessed at the end of the course.

Coursework

Fieldwork/coursework, in whatever guise, forms part of either AS or A2. There is no minimum weighting for coursework in Geography. Specifications vary between 20% and 40% over the two years of study.

Key skills

AS and A2 Geography specifications identify opportunities for developing and assessing key skills, where these are appropriate to the subject. The range of key skills is broad, incorporating Communication, ICT, Application of Number, Improving own learning and performance, Working with others and Problem solving. However, the only key skill component that has to be assessed through Geography is Communication.

Should you choose to pursue the AS key skills qualification it would be possible, by following your AS Geography course, to reach and demonstrate Level 3 (the necessary level of achievement), in Communication, Application of number and ICT, and conceivably this could come from your normal class activities and by formal testing.

The Key Skill AS is a worthwhile qualification and demonstrates your ability to put your ideas across to other people, collect data and use up-to-date technology. The Key Skills AS, like all other AS's, is worth half the UCAS score of the more advanced A2 qualification.

Different types of questions in A2 examinations

Extended prose and synoptic essay questions

Essays are a mainstay of the examination system, and make a prominent return in many of the new specifications. Essays in whatever guise can be the downfall for those who are unprepared.

The basic components of an A2 Geography essay answer are as follows

- There is a correct and appropriate response to command words.
- There is relevant content to the question put forward.
- There is an approach which is both confident and direct.
- Work is paragraphed, with one correct, relevant and strong theme running through the essay.
- There is a structure to the essay, guided by an appropriate plan (written on the paper that is handed in).
- Case studies are used but not overused, and these draw on a variety of scales (local, regional and global, if required).
- Where a comment on process is needed comment is offered.
- Balance is achieved between too much fact and the more discursive aspects of the essay.
- Diagrams and sketches are used that move the essay on.
- With regard to structure, a basic introduction is important (it should capture the examiner's interest). An 'expansion' of ideas in the middle section of the essay (the 'meat' of the essay) should appear next, followed by a conclusion. (It is important this is not a repeat or a summary of what's gone previously.)
- That you have allowed enough time to sit back and correct mistakes and omissions.

One might rightly expect there to be an incline in the complexity and type of language and command words used at A2. The lists below attempt to show how the step up is made between AS and A2; there are many other command words used.

AS	A2
• **Describe**: details of appearance and characteristics needed • **How**: process or mechanism recognition • **Why**: explain • **State**: briefly, perhaps one word • **Illustrate**: with examples and case studies	In addition to the words to the left it also uses: • **Examine**: that is, investigate in detail • **Criticise**: perhaps explores weaknesses in arguments • **Discuss**: considers both sides of an argument • **Justify**: an argument in support of a particular view • **Comment**: a balanced view or judgement

At A2 some specifications are using stimulus material to introduce candidates to the topic; others expect an essay to be produced from a 'cold' start. Most specifications expect A2 essays to take candidates about 45 minutes to complete.

Short and long structured questions (SLSQs)

Some of the examination boards are utilising these types of question in their A2 assessment suites. Such questions are sub-divided up with each section building on the last.

With these questions the single most important thing to do is to read the rubric (the instructions for the exam). Answer the correct number of questions, and aim to spend time commensurate with those parts of the questions that carry most marks. Fill the spaces left for your answers, avoiding re-runs of the question. You have to quickly focus your thoughts and ideas, but don't hurry your actual answers! Perhaps small plans in the margin would help this process, certainly underline or highlight key parts of the question stem and command words.

Many SLSQs come with extra information, like maps, graphs and tables. These are included to help you, they are not just page decoration; detail from them is absolutely vital for your answers.

Case studies are as important in structured papers as they are in essay papers. SLSQs, if compared to essay writing, are completely different in their demands and the technique required to successfully answer them. Practice is important, to both bring on your technique and to ensure that you have the time issue, outlined above, completely sorted out.

Ordnance Survey and SLSQs

Most boards use OS maps, though to a highly variable degree. Most seek to use them for analytical and interpretative purposes, usually as a 'tail' to SLSQs. In the past lots of map interpretation had been reduced to exercises in recognition and detection, a great shame as it is a core geographical skill. Those boards using OS skills at A2 utilise analysis and interpretation of the OS map to both extend and develop your geographical intellect. (Space prohibits the inclusion of OS extracts and exercises in this text, but what follows is an attempt to cover the approach and themes covered at A2 Level.)

On the whole maps are not chosen to spotlight 'classic' cultural landscapes, or to focus on specific areas or landforms. They are usually chosen to represent problem areas for which an interpretation can be offered on the basis of printed evidence from both the physical landscape and/or the rural and urban landscape.

Typically questions at A2 ask for a description, analysis or synthesis of evidence presented by the OS map, or any combination of these three.

Descriptive accounts usually involve looking at physical or cultural elements on the map. The features are described in terms of their relief, vegetative cover, or settlement and communications might also be described.

Analysis in which map evidence is systematically analysed at an elementary level and in relation to other mapped distributions.

Synthesis in which analyses are collated to form a statement or interpretation of the physical and cultural features observed.

Written coursework investigation/examinations (WCI)

WCIs first appeared in the last revision of Advanced Level Geography. Its popularity with many centres has ensured a place for it with a number of the boards at AS and A2.

Generally there is a selection of study area to be made (usually a human or physical choice). You will usually have access to pre-release material, which enables you to become familiar with the selected topic, the purpose of the study, the theory relevant to the study and some data related to the study and methods of data collection. It is also important to experience the topic firsthand through some fieldwork, and to apply this field information in the examination proper.

Question spotting should be avoided on this type of paper. It is only too easy to focus incorrectly on irrelevant sections; responses that are inappropriate to the set question gain few marks. It is also important to allocate time extremely carefully, writing to fulfil the requirements of the mark allocation, rather than concentrating on just one area. Incompleteness is to be avoided. The best responses are short and respond to the command words.

For many this type of approach works, but be warned it isn't an easy option, it entails as much work as the individual fieldwork study and is marked to the same standard!

In many specifications it is possible to take this route in Year 12 (AS) and to opt for a repeat or individual study in Year 13 (A2), if your first result is poor.

The investigative study/individual enquiry

Used in some form by all the boards, this personal piece of work requiring primary data collection is based on an issue, problem or question. An external moderator usually approves titles, and the finished work is of some 2500 to 4000 words depending on the specification you are following.

These investigations are a challenging task for you to undertake, but one which if small-scale and focused, local, accessible and topical can be extremely satisfying and enjoyable.

Perhaps the toughest part of the individual study will have been the selection of the topic/title. You would have been advised to look first at your specification, at past successful titles and importantly to talk to your teachers.

You would have read your board's specification for any special features that are required or needed by them in the individual study.

As the individual study is such an undertaking you would have planned and prepared a timetable over a longish period of time to ensure you bring the study to its proper conclusion. And by now, if an investigative study is required in Year 13, it should be nearly complete. You may even have completed a study in Year 12, for AS.

Exam technique

Links from AS

In order to study A2 Geography you will of course have studied Geography to AS Level. It is true that some topics covered at AS have developed ideas first encountered at GCSE, but many study areas will have been new and other areas will have developed ideas in a new or different way. A2 will extend and develop your knowledge and understanding still further.

It is likely that your teachers will be looking at A2 for students that have an interest in the environment and the world around them, and want to develop and refine fully their knowledge and understanding of the subject matter covered at AS, and to further their studies with the more advanced ideas, concepts and processes offered at A2. Students who are willing to explore ideas in an enquiring and lively way, and who can pass on and communicate findings and ideas effectively, are most likely to get the most out of Geography at A2.

What are examiners looking for?

All questions of whatever type seek to assess your appreciation and attitude to 'content', that is content, be it physical, human or regional in nature, of the specification you are involved with. There are certain common qualities that all boards look for in candidates.

1. A knowledge of facts, basic vocabulary, geographical concepts, processes and theories.
2. An ability to use information in an organised way, supported by appropriate case studies and examples.
3. The appreciation that all geography content is dynamic.
4. A range of skills understood and used in a range of geographical contexts.
5. An ability to comment and evaluate world issues and problems is paramount.

Quality of English is now integrated into the marking of papers. You must be clear and accurate in your use of English (i.e. spellings like desert, erosion, environment and vegetation must be correct).

Some dos and don'ts

Examiners then, are trying to assess the degree to which you can demonstrate as many of the qualities listed above as possible. Common problems encountered by examiners when they are marking examination work include (and these are the areas that you need to avoid!):

- Candidates spending too much time on one question or a part of a question.
- Candidates who let words like coast or river trigger an 'all I know' type of response, avoiding the focus of the question.
- Candidates who over-learn favourite topics in the hope they will appear in the examination. At A2, especially when synoptic issues are covered, you have to have an extensive knowledge of a range of topics. Narrowing your choice is dangerous.
- Candidates who don't plan answers, whether they are in response to synoptic questions or structured questions. Quick simple plans are a must, rewriting the essay title on your answer sheet can also help.

- Candidates who fail to respond to command words (see page 14).
- Candidates who attempt questions that lead to the snake pit: if you don't understand plastic deformation (glaciation), flocculation (soils) or fronto-genesis (atmosphere) avoid these questions!
- Candidates who don't respond to the vocabulary in the question, i.e. channels are different to valleys, glaciation is different to peri-glaciation, environmental hazards could mean pollution and/or biological and/or geomorphologic hazards, weathering is different to erosion, and so on...
- Candidates who have a limited reserve of case studies to draw upon. And those that choose case studies that obscure rather than illustrate points.
- Candidates who seem to be unaware of the ploys to reach the highest level or tier in a question.
- Candidates who fail to read the rubric of the examination, or fail to organise their time effectively enough over the whole examination. These candidates invariably show all the signs of panicking, i.e. questions unfinished and poor quality of language, uncorrected grammatical and spelling errors and major and avoidable omissions. It is important to re-read your completed examination!
- Candidates who have no real knowledge of their specification and how knowledge, understanding and skills fit each module/paper that they sit. Failure to practice past papers is also very obvious.
- Candidates who don't seek out and outline inter-relationships.
- Candidates who don't 'give' requested information, i.e. a named area within a city.
- Candidates who don't offer supporting diagrams and sketches.

Practice questions

This book is designed to help you get better results.

- Look at the candidates' answers (**they are real answers, from real candidates**), see if you understand where they have done well or where they have slipped-up.
- Try the exam practice question and then look at the answers.
- Make sure you understand why the answers given are correct.

If you perform well on the questions in this book you should do well in the examination. Remember that success in examinations is all about hard work, not luck.

What grade do you want?

The grade that is eventually awarded to you will depend upon the extent to which you have met the assessment objectives of the exam board you are studying with. Clearly, to gain the best possible mark you will have worked hard over the year. You will be highly determined and motivated. If you have identified weaknesses in your knowledge and understanding you will balance this by improving and building on your performance in other areas of the subject.

For a grade A

You will be a student who can:

- show a comprehensive knowledge of places, themes and environments
- understand how physical and human processes affect the above three areas
- show sound understanding of concepts, theories and principles
- understand a wide range of Geographical terminology
- understand how all of the above connect and be able to convey your understanding at a variety of scales.

For a grade C

You will be the student that produces sound, rather than competent answers. You may have some weakness in your understanding and knowledge and may be unsure of some terminology. You may synthesise and communicate your ideas and views less effectively than the A grade candidate.

- To improve you need to master and improve upon all of your weaknesses.
- You must prepare fully. Practice past questions.
- Hopefully, you will read around the subject a little more and keep up with current affairs.

For a grade E

You cannot afford to miss any marks! Even if you find the subject matter difficult to comprehend and would be content with an E, there are ways you can improve your prospects.

- Start by memorising the terminology of the subject.
- You must practice past questions. Being able to answer such questions even if it isn't in exam conditions is a great confidence-booster. On difficult questions try and answer the easier parts first, come back to the tougher parts later.
- It is likely that the areas of the subject that interest you the least are the areas where you experience the most difficulties, whether it be on a structured or essay paper. Such questions can be attempted and you will gain marks even if you only get part of the way through.

If you are working towards an A grade you need to keep at it and retain both your motivation and persistence. If you are not so fortunate, rest assured you will improve. Reading this text thoroughly, including the preliminary pages and the chapters themselves, and by completing the practice questions, will start this improvement process!

What marks do you need?

The table below shows how your average mark is translated.

average	80%	70%	60%	50%	40%
grade	A	B	C	D	E

Four steps to successful revision

Step 1: Understand

- Study the topic to be learned slowly. Make sure you understand the logic or important concepts.
- Mark up the text if necessary – underline, highlight and make notes.
- Re-read each paragraph slowly.

GO TO STEP 2

Step 2: Summarise

- Now make your own revision note summary:
  What is the main idea, theme or concept to be learned?
  What are the main points? How does the logic develop?
  Ask questions: Why? How? What next?
- Use bullet points, mind maps, patterned notes.
- Link ideas with mnemonics, mind maps, crazy stories.
- Note the title and date of the revision notes
  (e.g. Geography: Glacial activity, 3rd March).
- Organise your notes carefully and keep them in a file.

This is now in short-term memory. You will forget 80% of it if you do not go to Step 3.
GO TO STEP 3, but first take a 10 minute break.

Step 3: Memorise

- Take 25 minute learning 'bites' with 5 minute breaks.
- After each 5 minute break test yourself:
  Cover the original revision note summary
  Write down the main points
  Speak out loud (record on tape)
  Tell someone else
  Repeat many times.

The material is well on its way to long-term memory.
You will forget 40% if you do not do step 4. GO TO STEP 4

Step 4: Track/Review

- Create a Revision Diary (one A4 page per day).
- Make a revision plan for the topic, e.g. 1 day later, 1 week later, 1 month later.
- Record your revision in your Revision Diary, e.g.
  Geography: Glacial activity, 3rd March 25 minutes
  Geography: Glacial activity, 5th March 15 minutes
  Geography: Glacial activity, 3rd April 15 minutes
  ... and then at monthly intervals.

Chapter 1

Cold environments

The following topics are covered in this chapter:

- *Glacial activity*
- *The periglacial realm*

1.1 Glacial activity

After studying this section you should be able to understand:

- *that present day glaciated areas occupy a smaller area than during the Pleistocene*
- *how extremes of accumulation and melting affect the glacial system*
- *that glaciers can be classified and movement assessed and measured*
- *that glacial processes modify landscape, landforms and drainage systems*
- *that glacial processes both threaten and advantage human occupation*

LEARNING SUMMARY

Glaciology

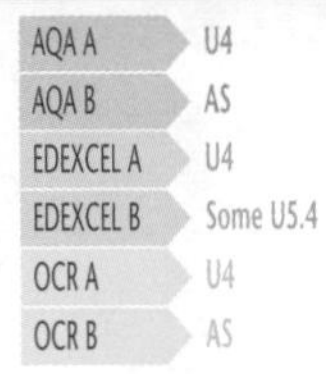

10 000 to 20 000 years ago, glaciers and ice sheets covered 32% of the land surface. Today 10% is covered by ice and 7% of the ocean surface is coated by pack- and sea-ice at the maximum winter extent. An additional 22% of the land is underlain by continuous or discontinuous zones of permanently frozen ground. The glacial age is not over, it has only diminished in its overall extent and intensity.

This period lasted two million years.

At maximum glaciation few areas of the earth were unaffected by ice activity and action. The last world-wide glaciation occurred during the Pleistocene epoch which comprised some twenty extreme cold periods (or major advances of ice) interspersed by periods of milder weather, **interglacials**.

Figure 1.1 *Maximum Pleistocene ice extent*

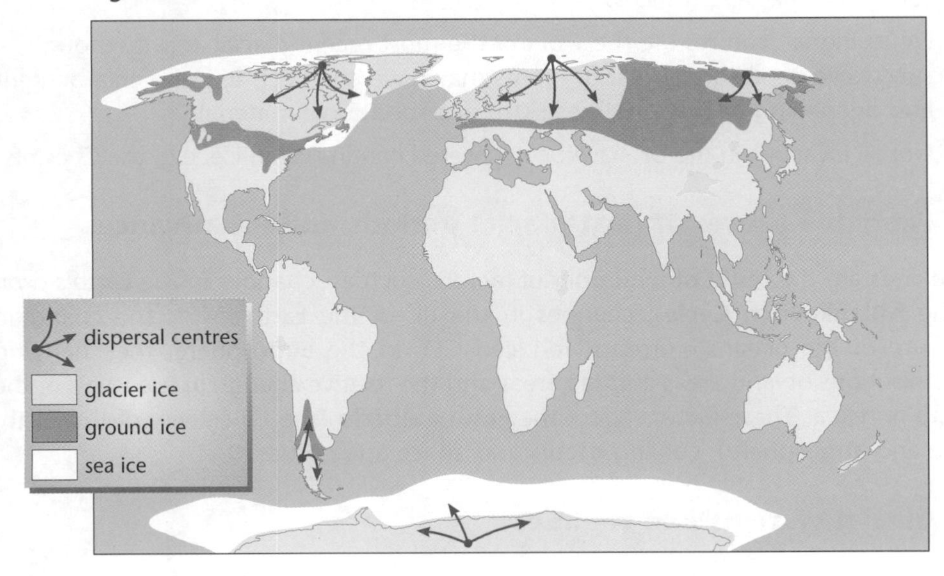

Figure 1.2 *Extent of glacial masses*

REGION	Present 000 km^2	Pleistocene maximum 000 km^2
Antarctica	12 588	13 200
Laurentia	153	13 790
Scandinavia	4	6 670
Asia	115	3 370
Rockies	77	2 500
Greenland	1 803	2 160
Others – N. Hemisphere	131	4 070
Others – S. Hemisphere	27	1 020
TOTAL	14 898	46 780

Ice Age Britain

Realise that most of the UK would not look like it does without the effects of glacial activity.

No area of the UK escaped the effects of the last ice age: the landscape we see today was created by ice advances over the British land mass. Three main advances have been recognised:

- The **Anglian advance** 420 000 to 380 000 years ago. This extended the furthest south, leaving major deposits in East Anglia.

Key points from AS

- **Erosion and deposition in rivers**
  Revise AS page 23
- **Coastal concerns**
  Revise AS page 43
- **Slope processes**
  Revise AS page 73
- **Soils**
  Revise AS page 83
- **Population density**
  Revise AS page 113
- **Isostacy**
  Revise AS page 44

- The **Woolstonian advance** 170 000 to 130 000 years ago. This originated in the Highlands of Scotland and affected most of Northern Britain, largely wiping out the effects of the Anglian advance.
- The **Devensian advance** Began about 115 000 years ago with a maximum advance about 30 000 to 18 000 years ago. Depositional and meltwater features in Shropshire, Cheshire and Yorkshire mark the limit of the advance. Ice left the land 14 000 to 10 300 years or so ago with the final brief advance from the Loch Lomond area (see diagram below).

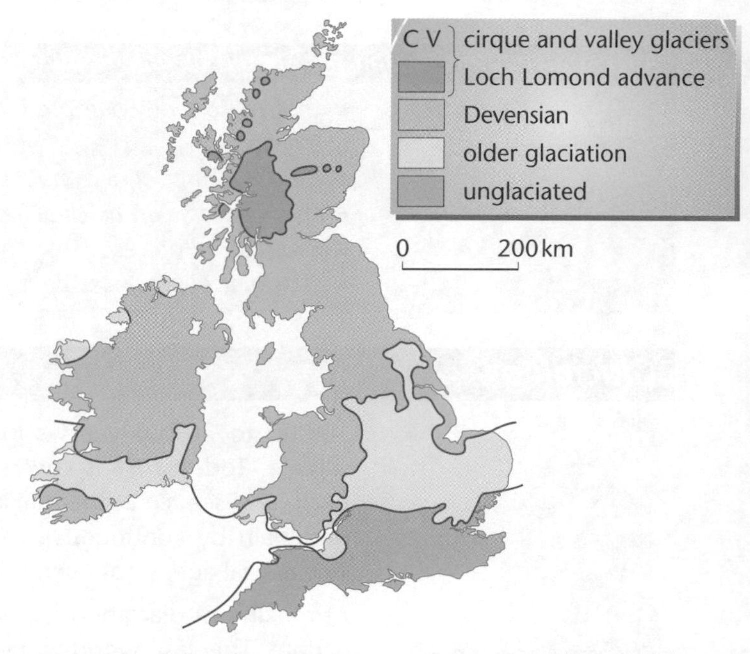

Figure 1.3
Limits of the main glacial advances

In summary, large areas of the UK have been heavily affected by glacial activity but three points should be emphasised:

- Erosional landforms are likely to have been influenced by multiple and successive glacial advances.
- Depositional features are the work of the most recent glacial activity though successive depositional/glacial events may have an effect, e.g. the enlargement and deepening of East Anglia's 150 m of depositional material.
- Not all locations in the UK have been covered continually in ice, e.g. the Cheviots.

Probable causes of past glacial periods and ice advances

Glacial periods are cyclic – be aware that we are probably in between another ice age!

Ice ages are the result of a number of factors, such as: changes in the Earth's orbit (the **Milankovitch Cycle**); changes in the tilt of the Earth's axis; the changing nature of the oceans' currents; reduced CO_2 in the atmosphere; the changing distributions of land areas and sea areas and the relative change in the levels of the land and sea. These factors affect the Earth's **albedo** (the reflectivity of the Earth, sea and atmosphere), cooling occurs and an ice age ensues.

Glacial systems and regimes

Glacial systems

A system is the balance between inputs, storage and outputs. Glacial systems attempt to maintain dynamic equilibrium (see figure 1.4).

Glacial regimes

The glacial regime describes the gain and loss of snow and ice in a glacier (or ice sheet). The regime determines the size of the glacier and whether it is advancing or retreating (see explanatory diagram figure 1.5).

Figure 1.4 The glacial system

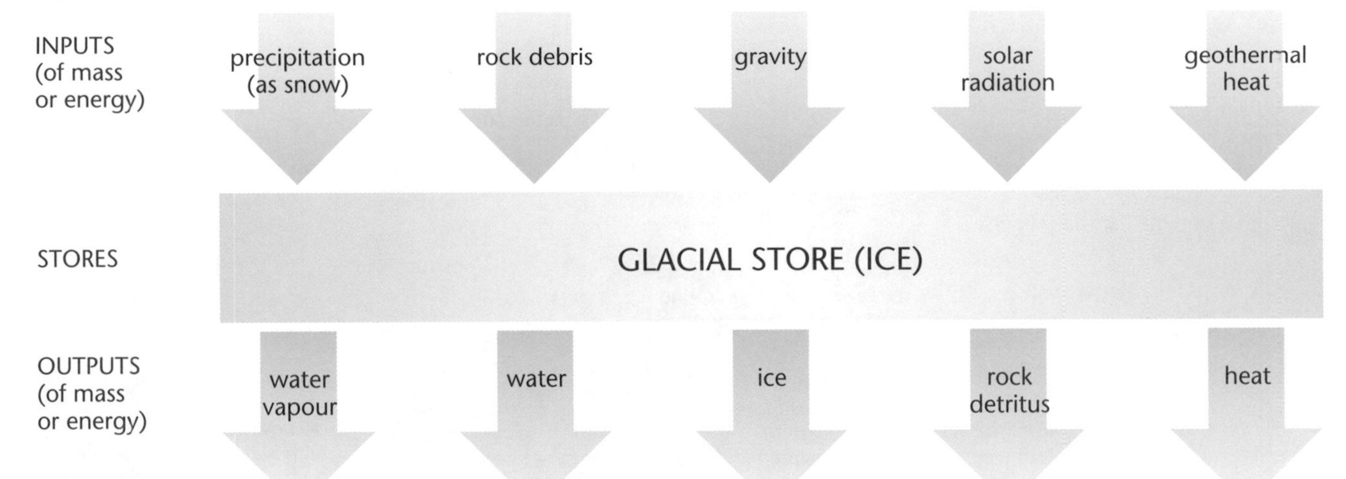

- **Accumulation** the gaining of snow (gaining of mass), is the input into the glacial system. It occurs high on the glacier or ice sheet.
- **Ablation** loss of snow/ice (loss of mass) is the output from the glacial system. It occurs near the margins or snout of the ice mass.
- **Positive regimes:** supply > loss by ablation. The ice thickens and advances.
- **Negative regimes:** supply < ablation. The ice thins and retreats.
- **Balanced regimes:** supply = loss. A steady state, the ice mass remains constant.
- The **glacial balance** or **budget** describes the annual fluctuation in ice mass position. Glaciers advance or grow in winter and retreat in summer, a pattern which can be shown diagrammatically/graphically, see figure 1.6.

Converting snow to ice

This can be expressed as a flow diagram.

Snow falls ⇒ snow accumulates (**alimentation**) ⇒ compression occurs ⇒ powder snow ⇒ **firn** or **névé** ('half ice') ⇒ pelletted ice with air spaces forms ⇒ the following season more snow falls ⇒ accumulates ⇒ compression occurs ⇒ flakes melt ⇒ water refreezes between previous seasons' pellets ⇒ white névé (density 0.06 gm/cm^3) gradually turns to blue glacial ice (density 0.9 gm/cm^3) as more air is driven out by compaction, re-crystallisation and re-freezing ⇒ the ice has reached some 30 m thick after several seasons of accumulation and compaction. Between the interlocking grains of snow and ice is an inter-granular layer of chlorides and other salts, these can lower the freezing point, thus ensuring some liquid remains in the system.

This helps with lubrication plus subsequent movement of ice masses.

Figure 1.5 Factors affecting mass balance

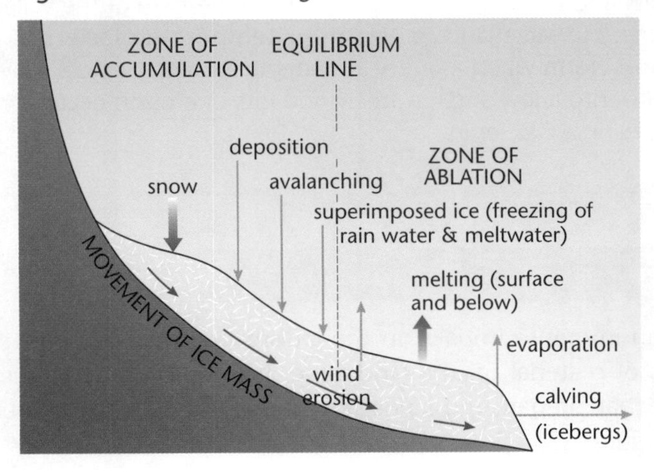

Figure 1.6 The glacial budget or balance

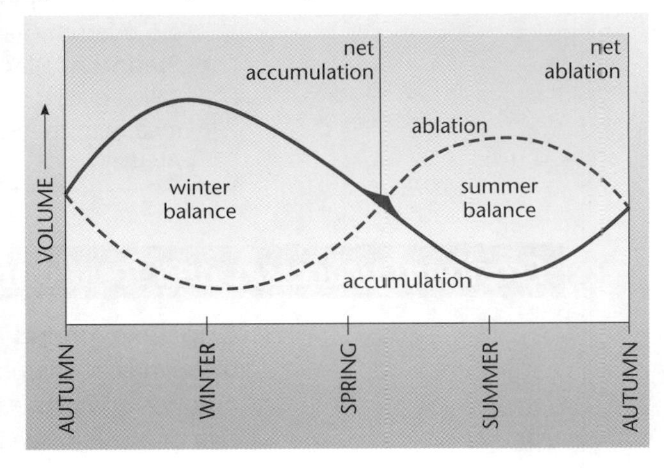

Figure 1.7 Slip and creep

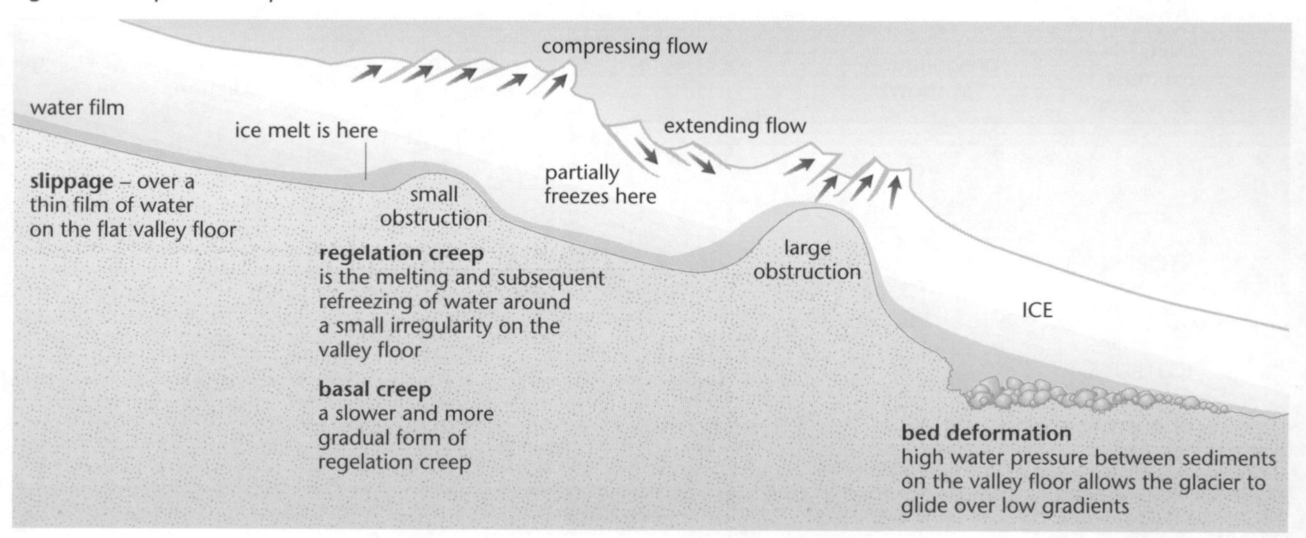

Pressure melting point, has the effect of melting ice. Warm ice moves more rapidly than cold i.e. The Black Rapids Glacier, Alaska moves at a speed twice as fast as the 7 km/annum of the Jacobshaun Isbrae ice-sheet in Greenland.

Friction holds ice back on the valley sides + trough floor.

Ice on the move

- Ice movement can be difficult to understand, as ice can behave as a plastic or stiff liquid. Whether the ice is plastic or near solid is controlled by the gradient of the 'rock floor', and the thickness of the ice (which controls the pressure melting point and temperatures within the ice). Smaller, thinner valley glaciers tend to move more rapidly than the massive ice sheets.

Actual types of movement include:

- **Basal slippage**, where a thin film of melt water allows a glacier to slide over the bedrock (see figure 1.7).
- **Internal deformation** where gradients are high. The glacier changes shape under its own weight, with ice particles moving relative to one another or in layers – a form of laminar flow.
- **Compressive flow** when compaction causes ice deformation on shallow slopes.
- **Extending flow** when thinning of the ice causes ice deformation on steep slopes.

Glacial classification

The classification of physical processes is a favourite at A2 Level. Know how glaciers and ice sheets can be classified.

KEY POINT

Size

- **Wedge or snow patch glaciers**
- **Cirque glaciers**, develop in hollows/basins high in the mountains, e.g. Cwm Idwal contained such a glacier
- **Valley glaciers**, form when a cirque glacier exits into a valley, e.g. Aletsch Glacier, in Switzerland
- **Piedmont glaciers**, form when a valley glacier spills onto a lowland area, e.g. the Malaspina Glacier in Alaska
- **Ice caps**, small areas of ice
- **Ice sheets**, large areas of ice, e.g. Antarctica

Thermal

- **Cold glaciers**, the ice is very hard and frozen to the base of the valley
- **Warm or temperate glaciers**, heat generates water, glaciers move freely and lots of erosion occurs

Erosional processes and landforms

AQA A	U4
AQA B	AS
EDEXCEL A	U4
OCR A	U4
OCR B	AS

Glacial erosion has an immense impact on the landscape, in some cases shaving thousands of metres of material off the landscape. It is believed that there are a number of variables that prepare the landscape and determine the amount and rate of erosion, see figure 1.8.

Figure 1.8 Erosional factors

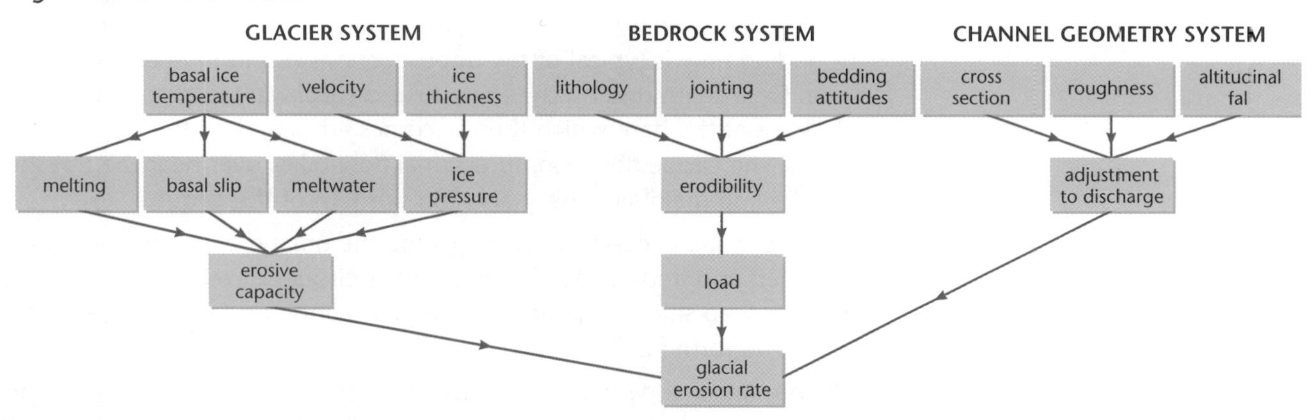

Indirect processes			Direct processes	
Surface rotting	*Pressure release*	*Plucking (quarrying)*	*Crushing & fracturing*	*Abrasion*
Frost action, weathering, mass movements, wind and water action in the vicinity of the ice mass.	Glacial advances and retreats constantly change the overburden pressures. Lots of material is made available for subsequent advances.	Water frozen into cracks and crevasses is ripped out as the glacier moves downstream. Plucking occurs mostly on the bed and sides of the glacial trough.	Due to the sheer weight of ice.	Debris carried by the glacier scrapes and scratches the glacial trough.

Erosional landforms

Effects vary from one area to another depending on relief. The Pleistocene can be seen in the present-day landscapes of Northern Europe and North America particularly in areas of high and varied relief.

Landforms produced by glacial erosion include:

Cirques, cwms or corries

Figure 1.9 Formation of a cirque

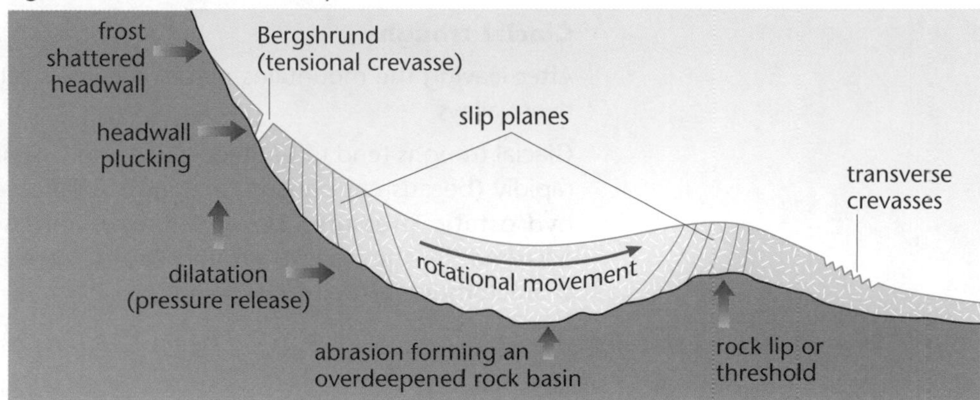

- These are large, rounded, armchair-shaped basins consisting of steep, frost-shattered backwalls, a smooth overdeepened floor and a moraine-covered threshold, or lip. They vary in size from the huge Antarctic cirques to the smaller British examples like Cwm Idwal in North Wales which has a backwall height of 420 m and an excavated area of 1.38 km^2. In general their length to height ratio is between 2.8:1 to 3.2:1 and shows a preferred orientation, onto shaded NE-facing slopes.
- **Nivation** is thought to be one of the main initiators of cirques. Snow accumulates in small, scattered hollows and, as the depth of snow increases, the hollows are gradually enlarged by frost shattering, meltwater erosion and perhaps chemical weathering too.

- Loaded with large amounts of debris and driven by the steep gradient and large inputs of snow, the cirque glacier starts to move by **rotational slip**. Extending flow is typical of the upper sections resulting in abrasion of the floor. Frost shattering on the **bergshrund** (backwall) leads to an accumulation of debris at the base which then becomes entrained by the ice and is used to abrade the cirque floor. Compressional flow takes over near the threshold or rock lip and maintains the shape and position of the cirque.
- As the cirque increases in size over time the upland area becomes heavily dissected and fretted. When two or more cirques develop close together a narrow steep sided ridge or **arête** (e.g. Striding and Swirral Edge, in the Lake District) is formed.
- When more than two cirques form close together a **pyramidal peak** or **horn** forms (e.g. Helvellyn in the Lake District, the Matterhorn in Switzerland). Both Helvellyn and Striding Edge were formed by the headwall retreat of Red Tarn and Nethermost Cove cirques.
- After glaciation the cirque often contains a small lake or tarn held back by the lip or threshold.

Be able to explain much of this in diagrams. At A2 a diagram can save a hundred words!

Figure 1.10 Cirque growth produces a 'fretted upland'

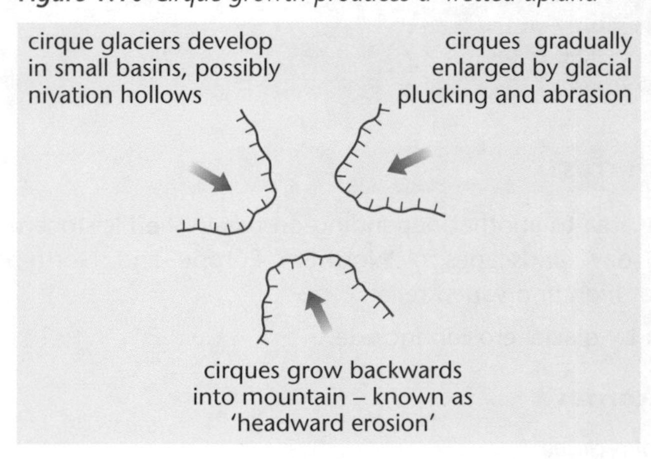

Figure 1.11 Postglacial 'dissected upland'

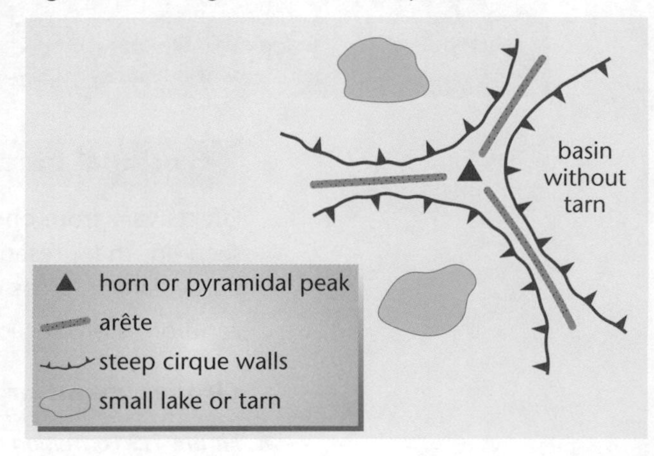

Glacial troughs

After leaving the mountains glacial activity is often concentrated into pre-existing river valleys.

Glacial troughs tend to be steep-sided and U-shaped. Thick ice in the valleys moves rapidly (because of pressure melting) aided by sub-glacial streams flowing under **hydrostatic pressure**. Down-cutting is intense, i.e. up to 600 m of erosion occurred in the valleys around Windemere. The diagrams below show the characteristics of the glacial trough.

Figure 1.12 Block view of trough

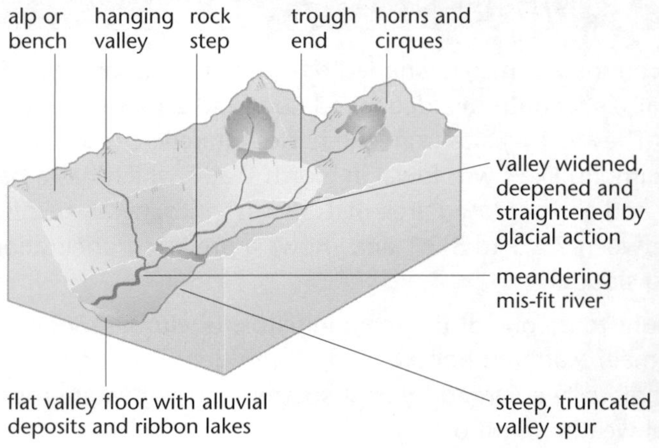

Figure 1.13 Long profile of trough, post glaciation

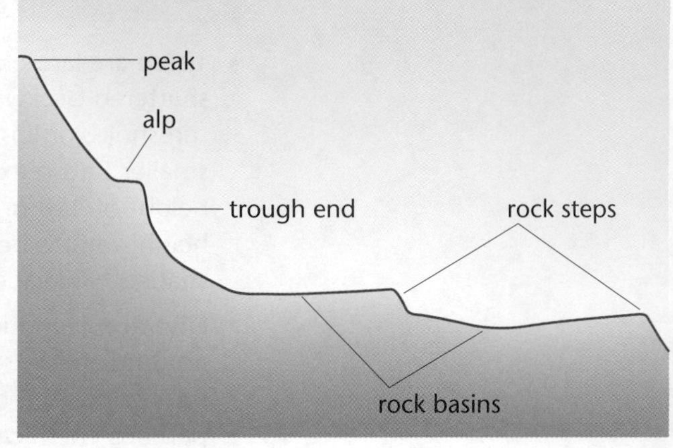

Figure 1.14 Roche moutonée formation

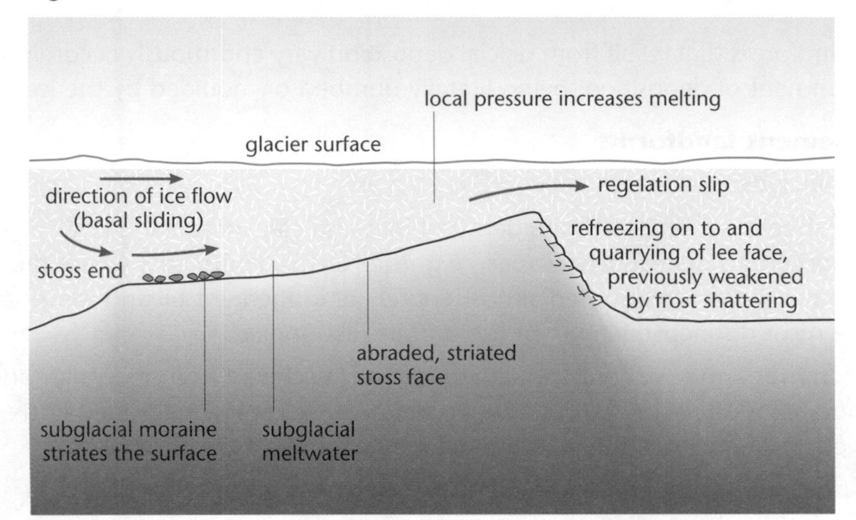

Figure 1.15 Crag and tail formation

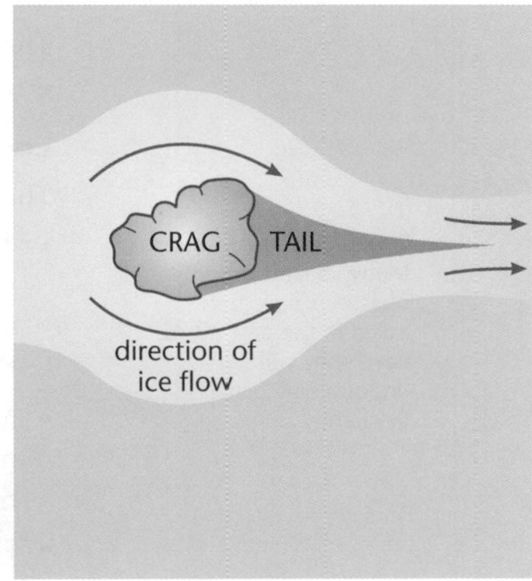

Rock lips and steps are thought to form because of changes in rock type.

- The head of the trough is marked by a sudden change of slope called a **trough end**; the long profile of the valley displays a series of rock steps and intervening basins that often contain ribbon lakes.
- The erosive effects of the ice ensure that tributary valleys are cut and subsequent drainage has to descend to the over-deepened valley, in the form of **hanging valleys**.

Other features of glacial erosion include **Roche Moutonée**, e.g. Grange in the Borrowdale Valley of the Lake District, and **Crag and Tail** formations, when resistant rock obstructs ice flow, and material is deposited on its leeward, e.g. Edinburgh Castle Rock.

Glacial diffluence

This occurs when the normal flow of a valley glacier is blocked. The glacier increases its depth and thickness and eventually flows across the lowest point available forming a **col** or **gap**, e.g. the Freshie, Geldie and Dee system in Scotland.

Glacial deposition and resulting landforms

AQA A	U4
AQA B	AS
EDEXCEL A	U4
OCR A	U4
OCR B	AS

Drift

The term drift refers to all types of deposits left after glacial ice has retreated. Drift is laid down in four distinct environments:

- **glacial** – dropped directly from the ice mass, unsorted and unstratified
- **fluvio-glacial** – laid down by meltwater streams, sorted and stratified
- **glacio-lacustrine** – lake deposits, sorted and stratified
- **glacio-marine** – deposits laid down into the sea, sorted and stratified.

East Anglian drift ranges in depth are from 143 m to 175 m.

Approximately 10% of the Earth's surface has a topping of drift; mostly lowland areas where the ice has slowed and ablation is high. Drift yield occurs in a variety of sizes and is variously smoothed and sorted depending on the environment it is collected in (upland deposition occurs whenever ice velocities drop).

Glacial deposition or **till deposits** are usually sub-divided into:

- **sub-glacial lodgement till** – associated with active glaciers, plastering the bed and sides of the valley with rock debris
- **supra-glacial till** – dumped onto the glacier, as moraine, finding its way, when the ice melts, onto the land surface
- **en-glacial material** – material within the ice is dumped by **melt-out** or ablation processes and tends to be well sorted and rounded.

Landforms

The landforms that result from glacial deposition vary enormously according to the environment of deposition; being literally dumped or moulded by the ice.

Lodgement landforms

These include:

Drift affects many areas. Many variants on drift questions have been asked in the past. At A2 Level they will undoubtedly occur frequently.

- **Till sheets** – a veneer of boulders, pebbles and clay, producing low monotonous plateau landscapes , e.g. the Norfolk landscape, where the 90 m Cromer Ridge is composed almost entirely of lodgement till and is probably the most distinctive feature of the drift in East Anglia.
- **Moraines** – material plucked, abraded and weathered from the valley and mountainside. It can be of end/terminal/push or recessional varieties. (N.B. medial, lateral and en-glacial moraines all contribute to this material.)
- **Fluted moraines** – form behind obstructions in/on the valley floor.
- **Drumlins** – rounded mounds formed parallel to ice flow, with broad upstream ends and a tapered downstream end. The streamlined elliptical shape causes minimum resistance to ice flow, with the broad end being the zone of maximum ice pressure. They are found in swarms or drumlin fields. They are between 150 to 1000 m long and about 100 to 500 m wide. Swarms form the so-called '**basket of eggs topography**' e.g. in the Afon Morwyn valley in Wales and the Eden valley in Cumbria.
  They may be formed in two ways: (a) by **deposition** – material dropped out of the glacier during melting is moulded into its stream-lined form by the retreating ice, (b) through **dilatancy** – stresses in till found near the base of moving ice masses deforms deposits during ice transport, causing the till to settle and compact, and the ice then flows around the mass of till, streamlining its shape.
- **Erratics** or **perched blocks** – rocks carried by moving ice are eventually dropped in areas where the dominant rock type is totally different. They are useful to geographers as they indicate the origins of ice and directions of movement, e.g. the Norber Block, North Yorkshire (see figure 1.16).

Figure 1.16
The Norber Perched Block

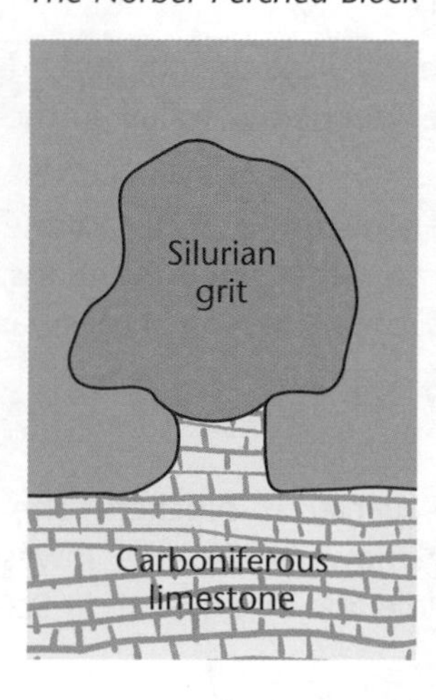

Meltwater or fluvio-glacial depositional landforms

Landforms produced by meltwater often have distinctive characteristics. They are smooth, well sorted/mixed and the water ensures the effects of glaciation are carried way beyond the ice front. The source of the material may be en-glacial, supra-glacial or sub-glacial; the water that carries the material appearing from sub-glacial channels, sourced from sub-glacial melting (during movement) and general ablation. The excessive load of the meltwater streams and the hydrostatic pressure, created by constriction, leads to choking of the water and channels that issue from the ice.

The Blakeney Esker is the best developed in the UK at up to 50 m high and 3.5 km long. In the same area kames can be recognised in the Glaven Valley.

- **Eskers** are ice contact deposits formed in meltwater streams in the very unstable sub-glacial environment of the glacier, e.g. Punkaharju in Finland.
- **Kames** form in the ice margins, deposited by heavily laden streams running into the glacier, e.g. Haweswater Trough in Cumbria. Sometimes these form terraces.
- **Kettleholes** – retreating ice often contains blocks of ice lodged in meltwater deposits. When this ice melts the local area collapses to form a kettlehole.
- **Sandar** is probably the most widespread meltwater deposit. This zone of deposition extends out well beyond the ice. The meltwater results in size sorting (known as grading) with the bigger particles being left near the ice front. There is a strong seasonal aspect to the production of this particular deposit. The huge load carried by the meltwater causes braiding to occur, e.g. the rivers of northern Iceland. Huge expanses of sandar have been spread over northern Europe forming sandy heathland, e.g. in north Norfolk at Kelling Heath.

Figure 1.17 During glaciation

surface streams
crevasses
deltas formed on lake bed
STAGNANT ICE
infilled lake
ice tunnels
marginal lake (partially infilled)

Figure 1.18 After glaciation

kame terraces
kames
kame terraces
eskers

Effects on landscape and land use

AQA A	U4
AQA B	AS
EDEXCEL A	U4
EDEXCEL B	Some U5.4
OCR A	U4
OCR B	AS/Some U5

Understand that past events affect present landscapes.

Glacial diversions and drainage modifications

The Avon and Thames are examples of rivers diverted from their original courses by glacial activity.

Many areas of lowland glaciated Britain were affected by meltwater from lakes blocked or impeded by lowland ice flows. Examples include Lake Lapworth in Shropshire, Lake Harrison in Northamptonshire and Lake Pickering in Yorkshire.

The **varves** (laminated clay deposits) deposited in these pro-glacial lakes have been variously settled and used for agriculture.

Figure 1.19 Proglacial lakes and overflow channels of the North York Moors

The valley of the River Esk was blocked by the Scottish Ice. Lake Eskdale overflowed through Newtondale Gorge (a typical spillway – flat floored and steep sided) into Lake Pickering. This in turn drained through the Kirkham Abbey Gorge. The present River Derwent rises within a few kilometres of the coast near Scarborough, cuts through Forge Valley (a lateral overflow), turns inland, flows through Kirkham Abbey Gorge (a direct overflow) and eventually reaches the sea 120 km from its source.

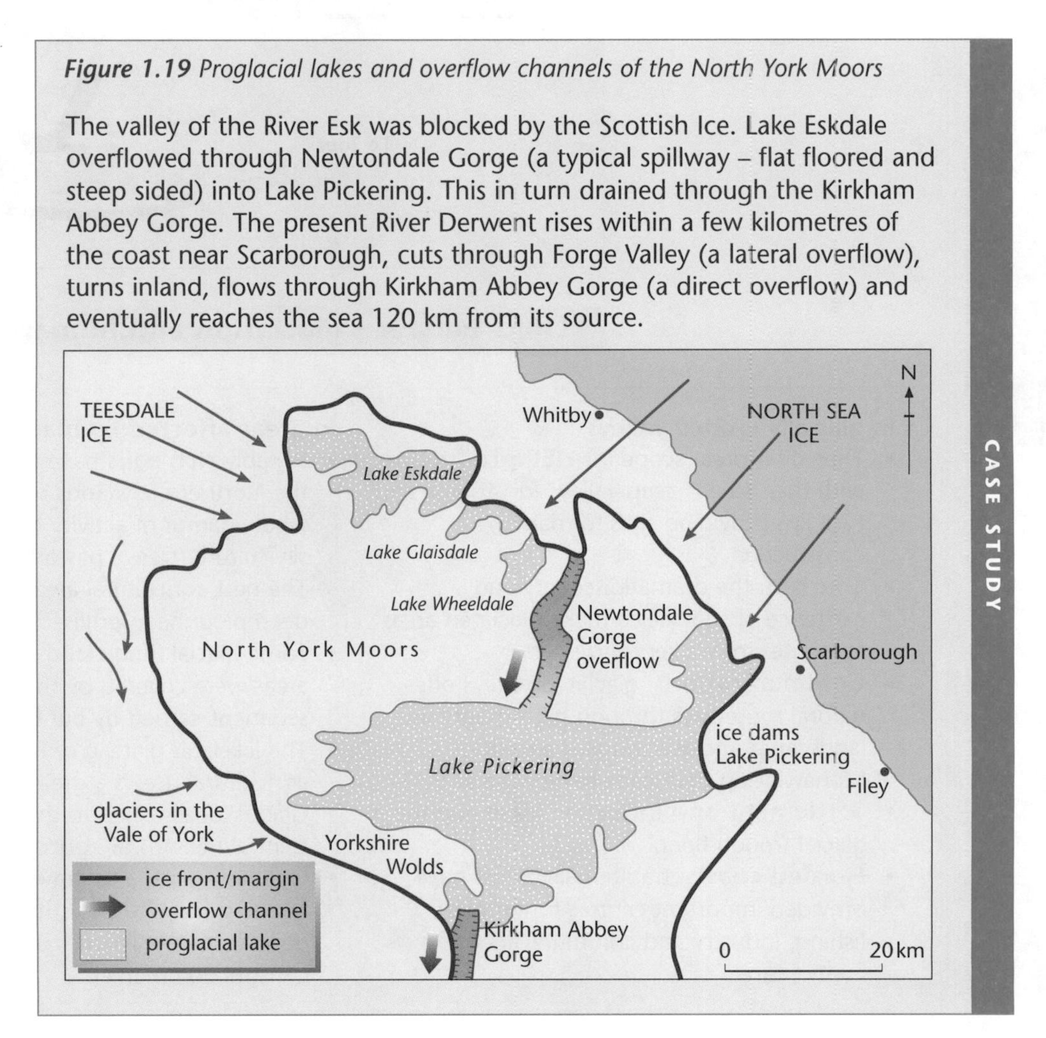

CASE STUDY

A present day glacial hazard: the avalanche

Figure 1.20 *Le Tour/Montroc avalanche, 1999*

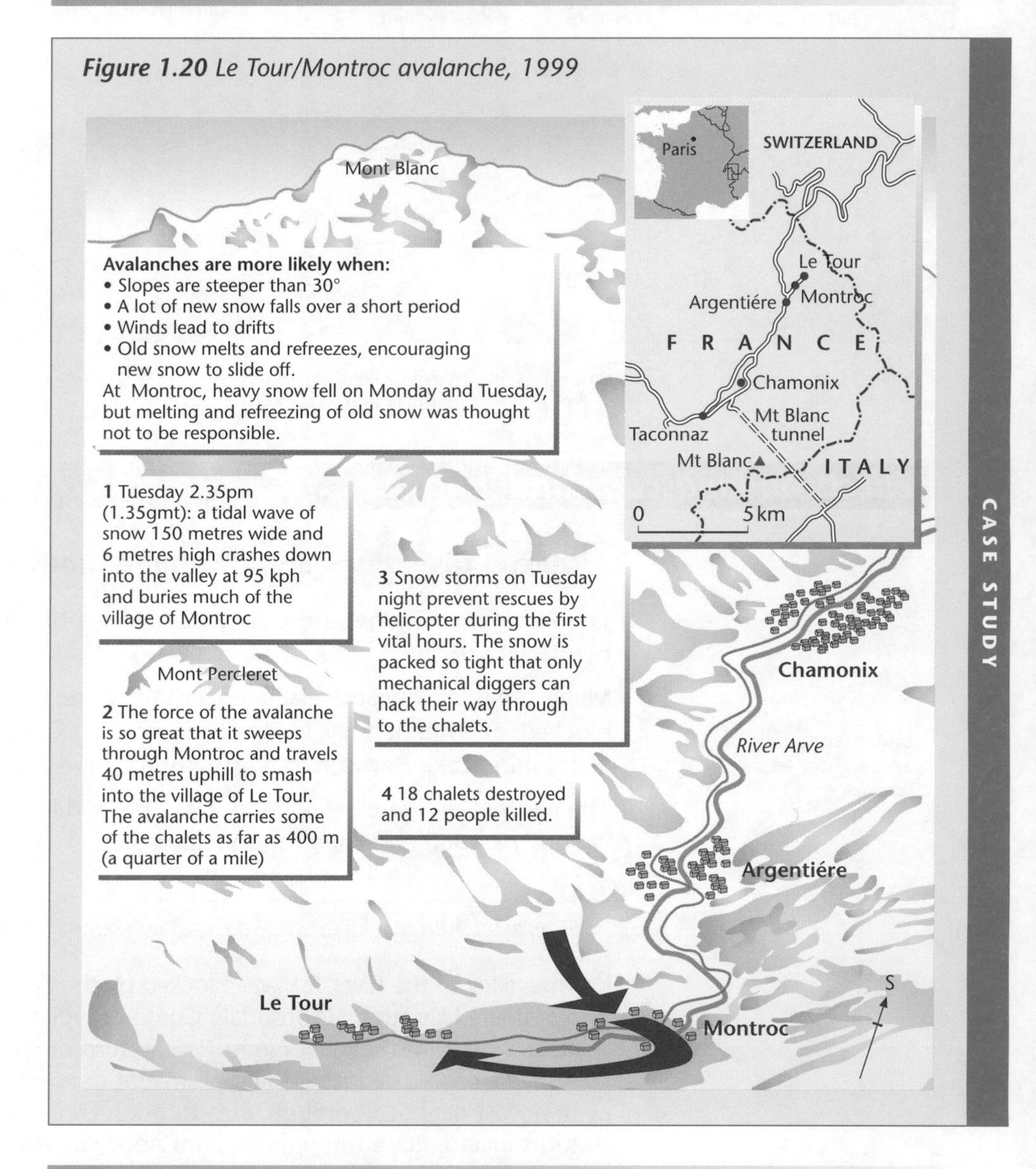

Links between glaciation and human activity

In glacially eroded valleys

- They offer great scope for **HEP production**; with their steep, deep valleys for storage and rock lips providing sites for dam construction.
- **Tourism**: the dramatic scenery and existence of snowfields make glaciated areas favourites for walkers and skiers.
- **Communications**: glaciated valleys offer natural routeways through high mountain landscapes and lower areas, e.g. the Mohawk Gap that leads to New York.
- **Settlement and industry**: these hug the glacial trough floor.
- **Fjorded coasts**: traditionally fjords have provided important centres for farming, fishing, industry and communications for many years.

In areas affected by glacial deposition

- Variably **rich soils** have formed in East Anglia and the Northern Lowlands of Europe. It has enabled a whole gamut of activity to be undertaken from silviculture (trees), pastoral and arable production. The best agricultural areas are those that have been deep ploughed, fertilised and irrigated.
- Some glacial fringe lands beyond the ice-covered areas were covered by thick layers of **loess** – fine sediment carried by out-blowing winds.
- The lakeland plateau of Finland, though of limited agricultural use, is a great **tourist attraction**.
- Glacial 'decline' led to the formation of peat bogs (in Central Ireland), **a source of fuel**.
- Outwash sands and gravels have been extensively quarried **for the building industry**, e.g. the deposits of the Thames Basin supply the vast London construction market.

KEY POINT

Gravel/sand quarries

Some have been utilised by local communities for leisure and recreation, others returned to nature.

Dorney Lake
Gravel extraction near the village of Dorney in Berkshire has created the 2000 m Dorney Lake, the new £15m Eton College Rowing lake. Gravel from the excavation is to be sold to off-set costs. The gravel will be used for a variety of projects in the Thames Basin.

The Cotswold Water Park
This is the UK's largest water park, 50% larger than the Norfolk Broads. It was created through gravel extraction in 132 pits.

Felmersham Gravel Pits
Gravel was quarried to build runways at local aerodromes during the War. Groundwater has filled the 52 acres of pits created, to form an important nature reserve near the River Great Ouse.

Whitlingham Country Park, Norwich
Created by the owners the Colman family, gravel and sand diggings are slowly creating new Broads, recreational lakes and a 2000 m rowing lake for the people of Norwich and East Anglia. Aggregates have been used for projects like the southern by-pass around Norwich.

CASE STUDY

1.2 The periglacial realm

After studying this section you should be able to understand that:

- *the areas presently affected by periglacial activity are dependent upon both present and past climate*
- *the active/mobile upper layer of the permafrost has an influence not only on the landforms that develop, but also on the use made of these areas today by man*

LEARNING SUMMARY

Characteristics of the periglacial realm

AQA A	U4
AQA B	AS
EDEXCEL A	U4
EDEXCEL B	Some U5.4
OCR A	U4
OCR B	AS/Some U5

Periglacial environments are found on the edges of ice-masses, that have permafrost conditions virtually all year around. Other characteristics include a summer thaw and lots of freeze-thaw activity.

In the frozen permafrost areas the top soil is known as the **active layer** (this is a highly mobile zone and one which experiences lots of mass movement, as it thaws out in the summer and re-freezes in the winter, see figure 1.2).

Some unfrozen ground **talik**, may exist within the permanently frozen sub-soil. Present day northern hemisphere permafrost is distributed as in figure 1.22.

The cyclic nature of the climate (**tundra climate**) and the freezing and thawing of the ground in these areas has led to a peculiar set of geomorphological processes to operate and landscape features to form.

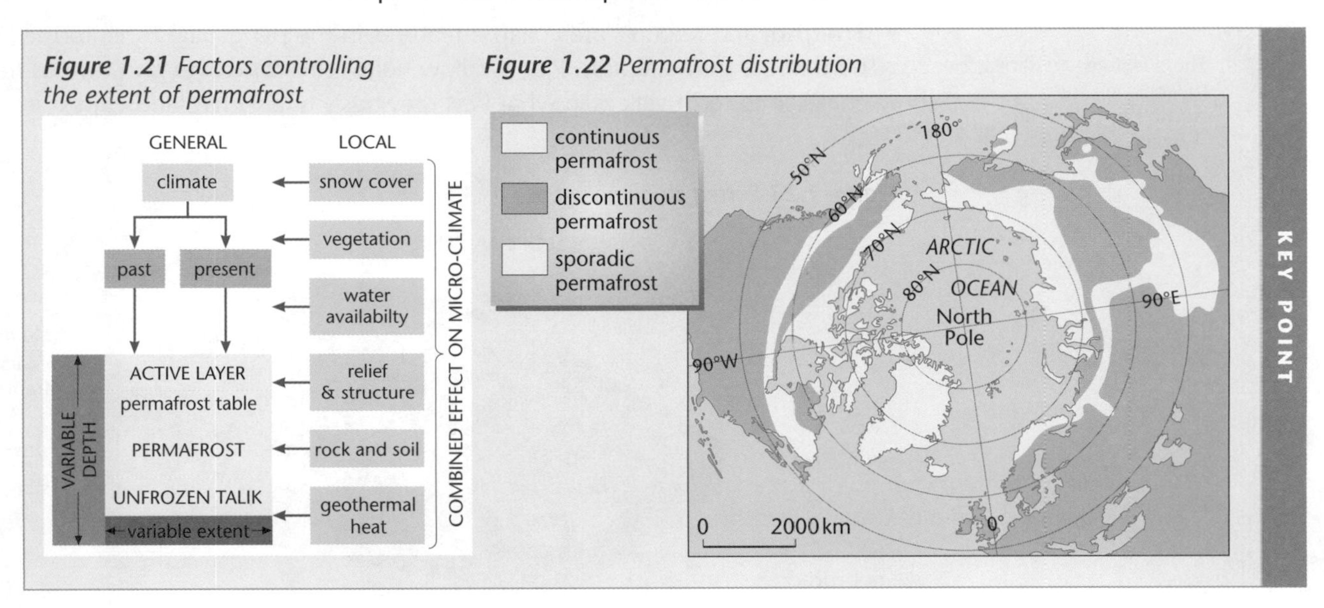

Figure 1.21 Factors controlling the extent of permafrost

Figure 1.22 Permafrost distribution

KEY POINT

Processes and landforms

Four major sets of processes/features have been recognised in these areas.

Processes creating new deposits

The periglacial realm has had a huge influence on our landscape and still affects polar regions.

- **Frost shattering** or **nivation** (the mechanical splitting of rocks by ice) is most active where there is a maximum number of temperature oscillations above and below freezing point. The result is thick layers of frost-shattered angular fragments, that cover huge areas. Those with the largest chunks are known as **blockfields**.

Periglacial A2 questions always look to relate man and his environment.

- **Talus** or **scree slopes** may develop as a consequence of this frost shattering activity. Once formed this loose material can become very mobile and is devoid of vegetation. The finer of the deposits constitute a mobile layer on the shallowest of slopes (<2°) of the permafrost/periglacial realm.
- In the warm summer, lubricated thaw mass movements called **solifluction/ gelifluction** move rubble around the landscape in lobes and trails.
- Wind-blown deposits known as **loess** also blanket such areas (**limon** in France and **brickearth** in Britain), a consequence of the strong out-blowing winds that are known to have dominated in such areas during the last ice age.

Structural features

Stone polygons vary in size from 0.5 m to 15 m in diameter. Stone wedges are 3–4 m deep.

- In the flat badly drained periglacial areas there is no continuous vegetation and lateral sorting of the active layer takes place to give a highly distinctive ground pattern of **stone polygons**. These are caused by contraction and expansion, a consequence of frost heaving, see figure 1.23.
- On steeper slopes the rings can become elliptical or elongated into **stone stripes**.
- Where cracks exist on the surface they fill with 'thaw' water, wind-blown sand and stones. An **ice wedge** forms that grows wider and deeper as each winter freeze and summer thaw cycle progresses. This process can lead to a very domed and hummocky landscape, as the ground expands against the ice wedges, distorting surface deposits.
- **Pingos** are mounds or cones covered with deeply fissured drift layers drawn into a mound by an intrusion of ice. They grow very slightly during their life cycle by drawing water to the ice mass, from the summer supply of groundwater. These features have been well studied in the Mackenzie Delta in Canada. In Britain the scars of pingos are all too obvious in the landscape of Breckland, Norfolk.
- **Thermokarst** features are collapse features in the periglacial zone, formed when permafrost thaws rapidly. Pits, swallow holes, caverns, caves and ravines may coalesce to form valleys in what was previously a fairly flat landscape.

These features are similar to limestone 'karst' landscapes (c.f. **dolines** ⇒ **uvalas** ⇒ **poljes**).

Figure 1.23 *Sorted stone polygons*

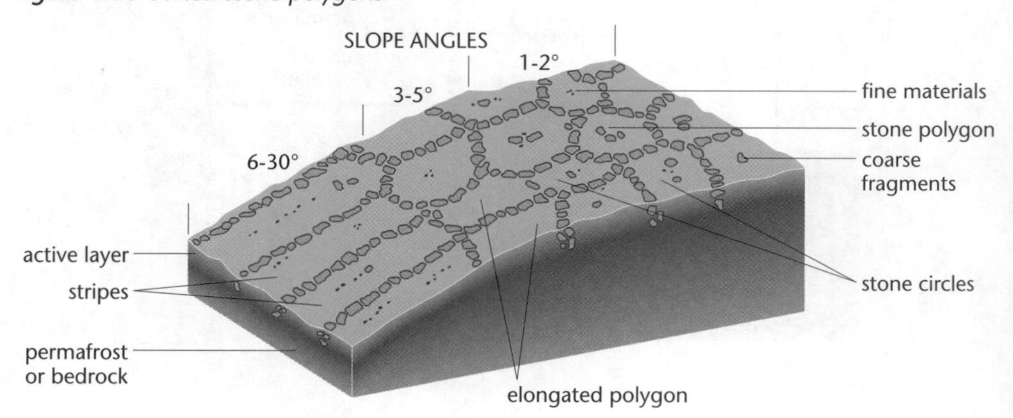

Rock lips are replaced by **protalus ramparts** in nivation hollows.

Landform modification

- **Angular free faces** are common, and **tor-like forms** are found where the joint structures intersect at right angles to the ground surface.
- **Altiplanation terraces** occur, benches cut into a hillside by the combination of frost shattering, nivation and solifluction.
- Small **nivation hollows** (similar to the cirques of highland glaciated areas) form on the sheltered north-facing slopes of the permafrost/periglacial area. The snow that occupies these hollows both deepens and extends them.
- On the south-facing warmer slopes disintegration processes and mass movement (solifluction) leads to a decline in slope angle. This leads to the formation of **asymmetrical valleys.**
- There may be evidence of river activity in the periglacial landscape. For instance, in the UK, erosion is seen in the form of **dry valleys** and deposition in the **coombe deposits** left in the bottom of the dry valley.

Figure 1.24 *The periglacial landscape*

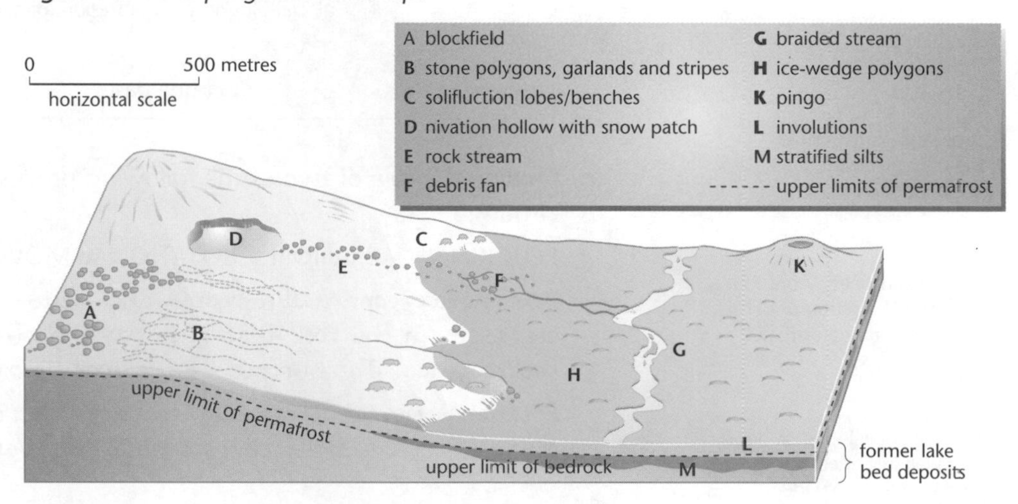

In Britain we can view the Pleistocene landscape of periglaciation in much of Southern England, see diagram below:

Figure 1.25 The periglacial legacy in Britain

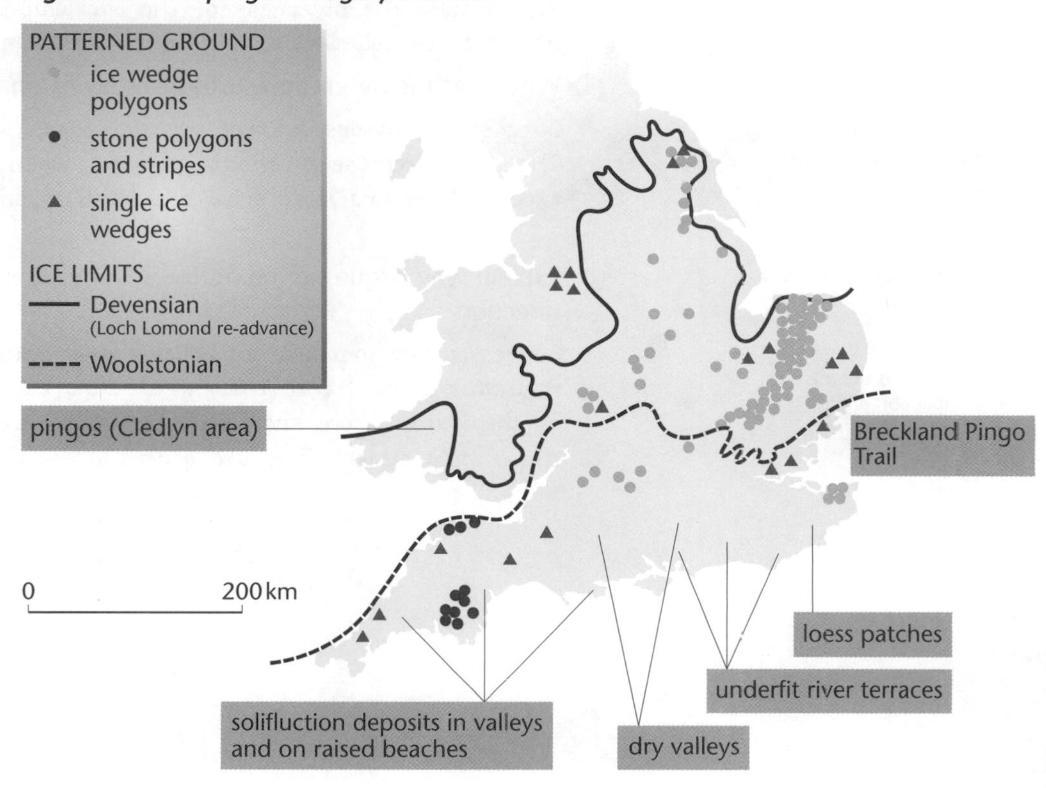

Sample question and model answer

1 Study the diagram, which is a cross section of a glacial cirque or corrie.

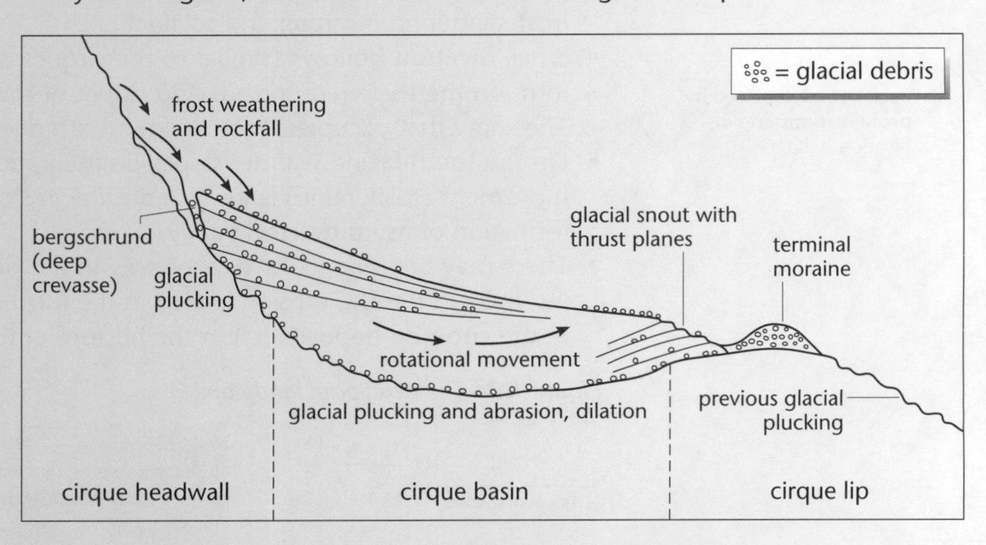

(a) Outline the role of **two** of the following processes in the development of cirques:

PLUCKING ABRASION ROTATIONAL MOVEMENT [4]

Plucking-ice continually freezes and melts around the rock and into crevasses in the rock that it is passing. This has a ripping effect on the rockface. The surface of the rock becomes very jagged and rough.

Rotational movement actually describes the movement of ice in the glacier. It causes the characteristic basin/armchair shape of the cirque.

(b)

(i) What contribution does the bergschrund make to corrie formation? [2]

Meltwater finds its way into the bergschrund. This cyclically freezes and thaws; the steepness of the backwall is maintained. It provides debris to the glacier to help in the abrasion of the corrie.

(ii) Why does the debris appear to be in layers on the glacier? [2]

Successive seasons are not necessarily as warm or cold as one another. The layers represent periods, perhaps, when accumulations have exceeded melting: i.e. snow has built up on previously deposited debris.

(c) Explain why cirques in the northern hemisphere tend to face in a northerly direction. [5]

Corries/cirques originate in hollows and cracks in the upland areas of mountains. The freeze-thaw weathering experienced increases the size of the hollow, more snow collects, more weathering occurs and the cirque gets bigger. They are generally found on N.E. facing slopes.

It pays to learn simple definitions – even at A2 they are needed!

The third process, Abrasion, is the scouring action of debris in the ice. It striates rocks.

Logically developed. A good and well informed answer. Understands the wide contribution the bergschrund makes.

Maintenance of the steep backwall is called **basal sapping**.

A good answer and one that most A2 students should be able to produce.

Alternative names: corrie, cirque, cwm.

Understand the ideas of compaction and accumulation.

Sample questions and model answers (continued)

But why? Its to do with frequency and intensity of summer melting and aspect.

Perhaps this answer needed a little plan – the important use made of such areas for communications has been missed.

A reasonable examination of this question understands the negative aspects.

Some exemplar support offered. But limited.

Dwells on the negative. Farming as a positive effect might have been offered (ditto forestry).

(d) Using examples examine the impact of humans in glaciated upland areas. [7]

There are few places that have not been touched by man. The deep valleys of the upland areas offer superb sites for HEP production, but bring with them the paraphernalia of power production. Wires, buildings and scars on the landscape. On our field course to West Wales we visited the Rheidol Valley and its HEP plant. For the most part it is well camouflaged but the plant itself and various 'collecting' reservoirs are only too obvious in the landscape.

The dramatic scenery and existence of snowfields make glaciated areas favourites with walkers and skiers – both activities intrude on the landscape causing footpath erosion and litter, and damage irrecoverably the vegetation of such areas.

Assessment and Qualifications Alliance

Comments:

The best answers were to be found in a and b. C is a little bit awkward and threw the candidate. D needed a plan. This was awarded 4 + 4 + 2 + 5 = 15/20

Practice examination questions

1 Study the article and answer the essay below.
It remains extremely difficult to develop and settle the cold environments of the world in a sustainable manner. Discuss. [25 Marks]

Russian villagers eat dogs in big freeze

by MARK FRANCHETTI

TENS of thousands of Russians are stranded in freezing temperatures in remote villages in Russia's far north without heat, running water or food. People have been forced to eat dogs. 50,000 people are struggling to survive in areas that have been completely cut off. In Oymyaakonsky Ulus in Yakutia, a vast region more than 4,000 miles northeast of Moscow, some 4,500 people could die unless they are rescued from a cluster of isolated gold-mining villages. There is no fuel and hardly any food left.

The region, which is stalked by packs of wolves, is the coldest permanently inhabited place on earth. Winter temperatures regularly plunge below –60°C.

A wrangle over whether the local or federal government should pay for the operation meant that only four villages were evacuated.

The only escape route now for the people of Nelkan is across Russia's vast, inhospitable north.

Keeping Russia's far north and far east alive has always been an enormous task; the region of Yakutia alone is five times the size of France. Stretching over several thousand miles, separated by 11 time zones and strangled by months of darkness and sub-zero temperatures, the northern regions were inhabited only by indigenous peoples until the Soviet Union unearthed their wealth of diamonds, gold, platinum, coal, oil and nickel.

Political prisoners were used as slave labour to extract the riches. About 12m people now live in the country's northern and far eastern regions; but keeping so many alive in a place where nothing grows and everything must be imported is the cost that the state can no longer afford. This year the Russian government allocated $70m to transport goods to these remote regions. According to Misnik, it needs 20 times that amount.

2 The diagram shows the accumulation and ablation of ice (the glacier mass-balance) along the length of a valley glacier.

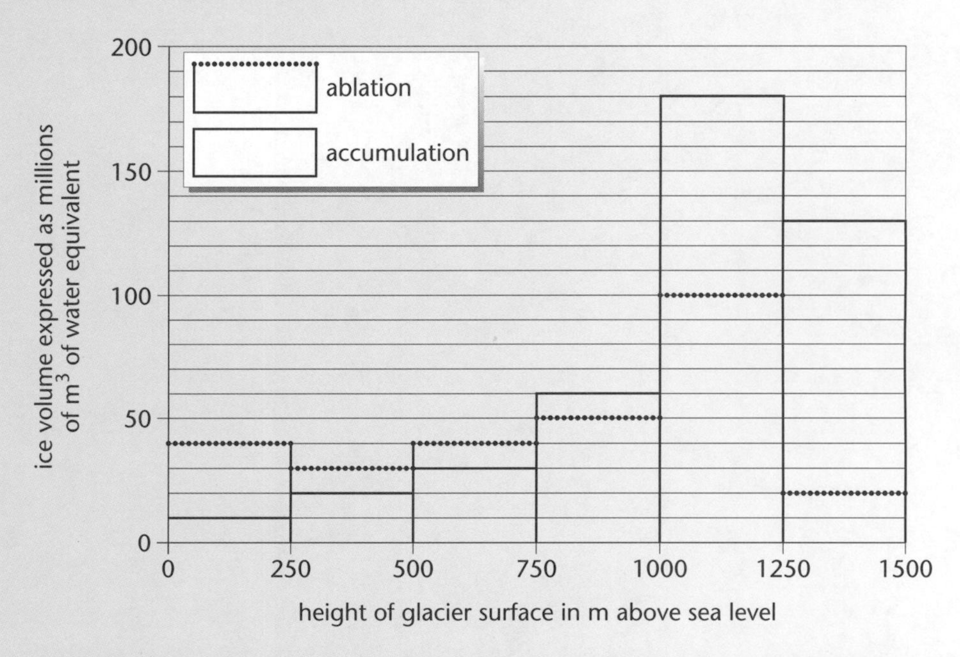

(a) How might the glacial mass balance of the glacier in the diagram affect its movement? [5]

(b) Explain how the movement of a glacier produces distinctive landforms. [20]

Edexcel

Chapter 2

Tropical environments

The following topics are covered in this chapter:

- *Tropical rainforest – TRF*
- *Tropical grassland – savanna*

2.1 Tropical rainforest – TRF

After studying this section you should be able to understand:

- *the global significance of TRF*
- *that TRFs are species rich and highly stratified*
- *the climatic features in the tropics – based on high rainfall and temperature*
- *the destruction versus sustainable development debate*

LEARNING SUMMARY

Distribution and characteristics

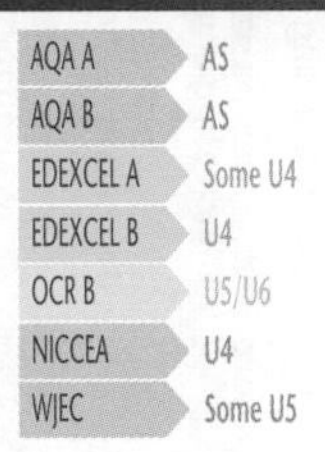

AQA A	AS
AQA B	AS
EDEXCEL A	Some U4
EDEXCEL B	U4
OCR B	U5/U6
NICCEA	U4
WJEC	Some U5

This complex and rich ecosystem contains the Earth's biggest gene pool, thought to be the result of mutations and the subsequent evolution of species because of the vast amounts of UV-B radiation absorbed. It is also the world's most prodigious terrestrial ecosystem, primary production/productivity being in the order of 1000 to 3500 g/m²/yr (the result of high rainfall and temperatures). Additionally the TRF plays a significant role in maintaining stability in the atmosphere. The TRF is the world's most finely balanced system.

TRF has a world biomass productivity of 765 000 tonnes/ha.

Distribution

The main areas include:

- the Amazon Basin and into Central America – 58% total area
- the Zaire (Congo) Basin in Africa – 19% total area
- the Indo-Malay/Indonesia to Australia – 23% total area

TRF is the world's oldest major ecosystem.

All these areas lie between 10° N and S of the equator; covering 718 million ha or 5% of the land surface of the Earth, see the map below.

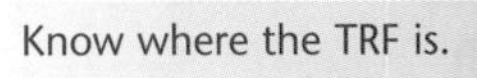

Figure 2.1 The global distribution of TRF

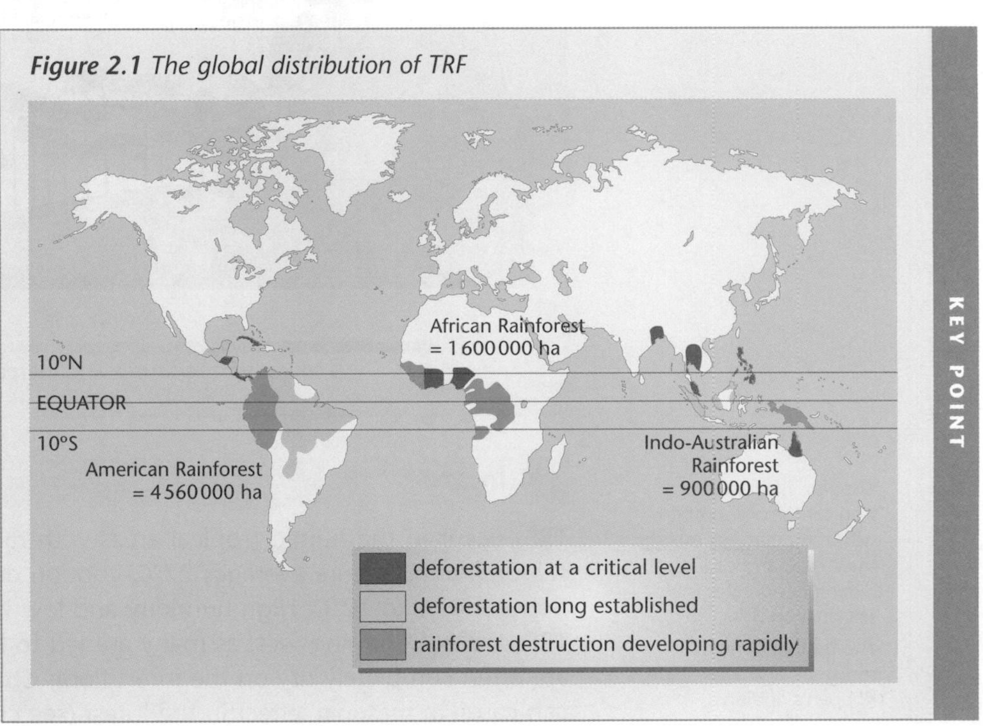

KEY POINT

Key points from AS

- **Global air movements** *Revise AS page 47*
- **El Niño** *Revise AS page 48*
- **Soil classification** *Revise AS page 80*
- **Global management of forests** *Revise AS page 86*
- **Optimum population** *Revise AS page 112*
- **Population change** *Revise AS page 110*

Plants which exist in the TRF, do not, as is generally thought, grow continually.

Vegetation

TRFs occur in similar climatic zones: they are the **climatic climax community** (see below). All TRFs resemble one another in structure, habitats etc. because they experience the same or similar environmental variables; in ecological terms they are a **convergent community**. Variations reflect differences in climate, relief and altitude, and **edaphic** (soil) conditions (see below).

***Figure 2.2** Vegetation layers in tropical forest*

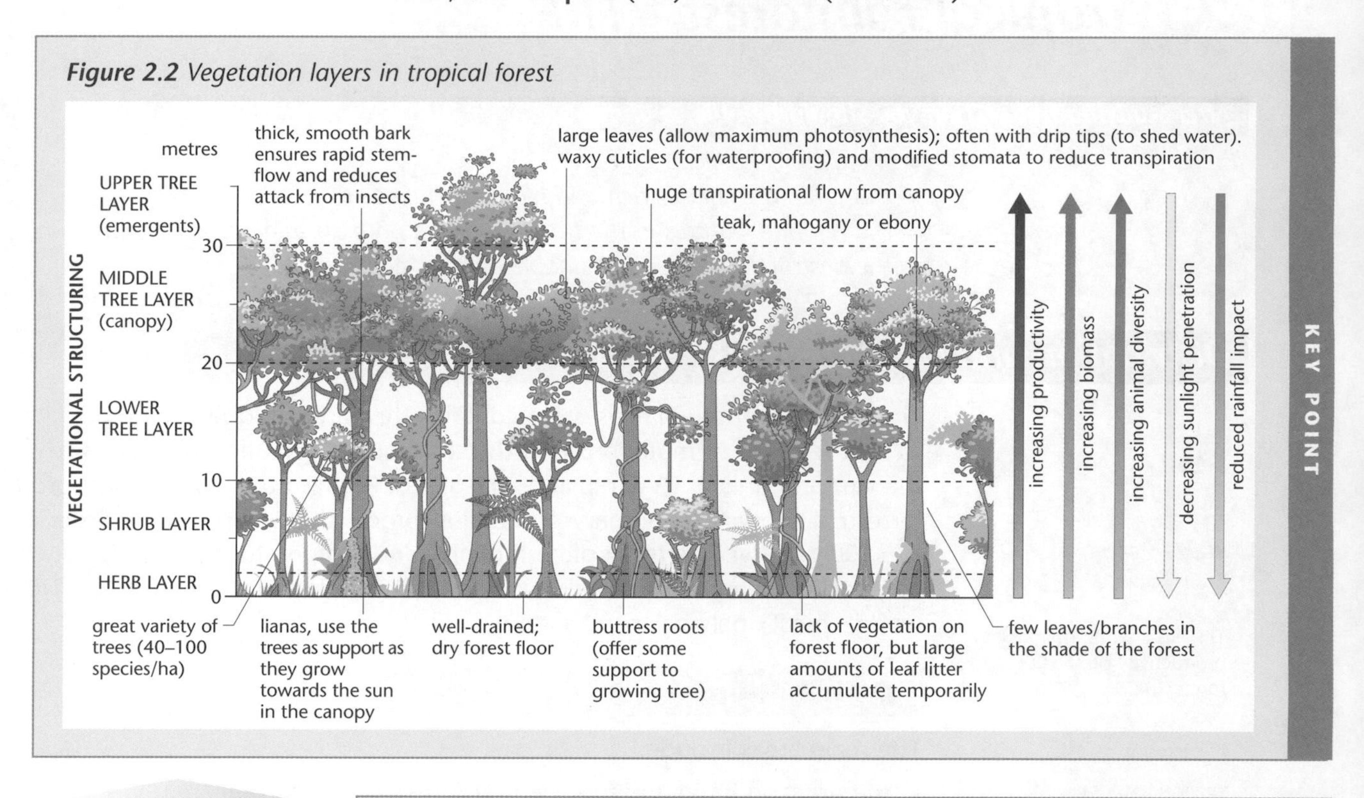

Students should be able to describe ecosystems in detail. With reference to vegetation, climate, and soil.

***Figure 2.3** The variants of the TRF based on relief, altitude and soil*

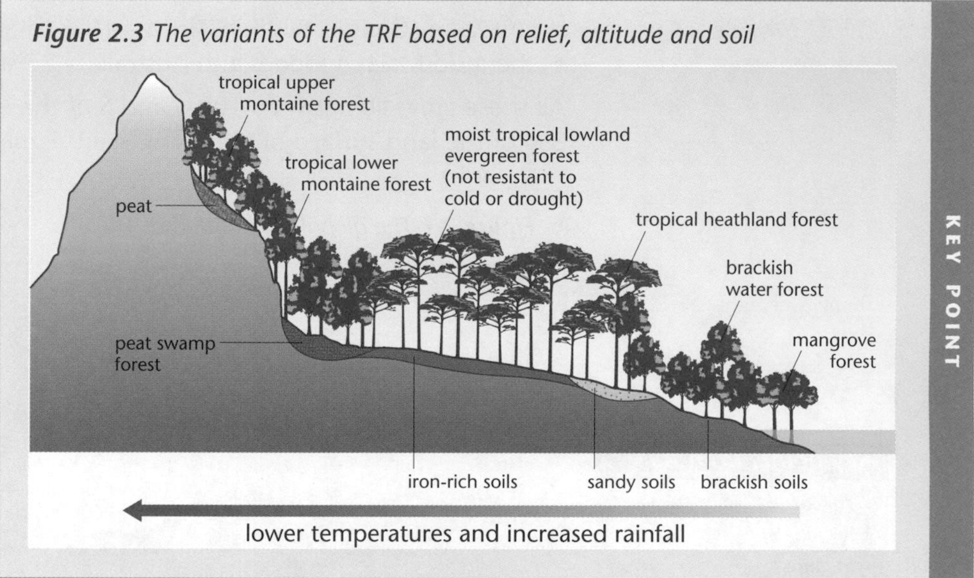

Koppen, identified a tropical climate as one with temperatures greater than 18°C and daily differences greater than seasonal ones.

The best students will recognise and appreciate Koppen's findings.

Climate

TRFs occur in the humid tropical areas with rainfall exceeding 2000 mm/yr. The seasonal temperature averages 27°C, though daily temperature swings can range between 6°C to 12°C. High humidity and low windspeeds are typical features but TRFs are not dripping wet, as many are led to believe. They are well drained and are often completely dry on the forest floor. Changes in altitude and microclimate can cause a 'seasonal' effect in the forest (see climatic graph opposite).

Fig 2.4 Climate in the TRF

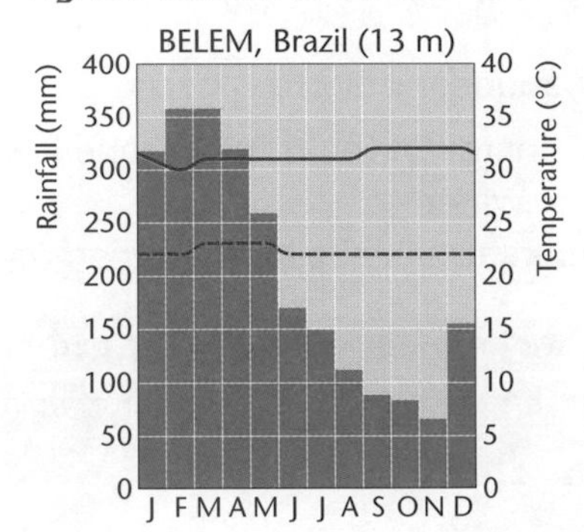

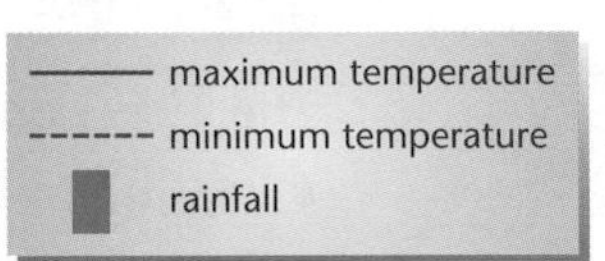

At A2 Level you should realise there is a rhythm to the weather.

***Figure 2.5** The daily rhythm of weather in TRF areas*

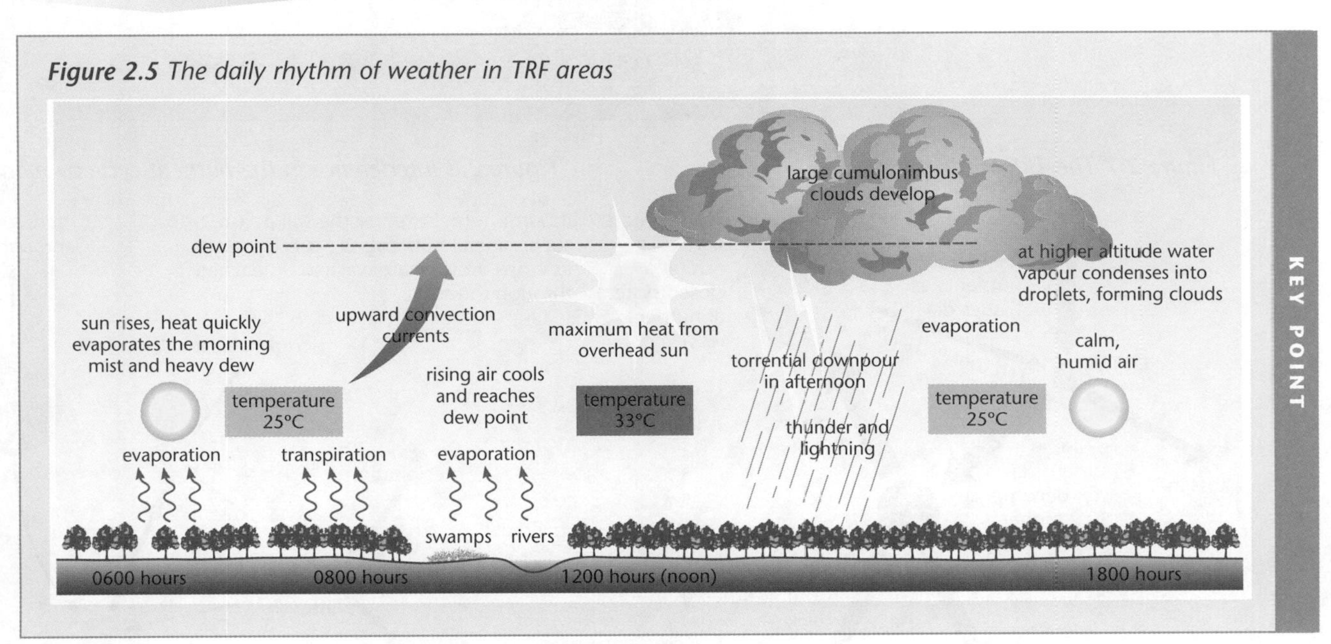

The tropical latosol soil

Typically these soils are:

- deep; up to 30 m
- very well weathered (in places turned into kaolinite, a type of clay)
- acidic and intensely leached
- rich in iron and aluminium in the upper layers; hence their red colour
- low in organic matter because of high rotting rates; they are humus poor
- downward drained, typically the soils receive more precipitation than is evaporated
- highly translocated; lots of material is moved through them by water, a process known as **ferrallitisation** in TRF areas
- highly leached; this is a dominant process.

The best candidates will always show an understanding of soil processes and realise the results of even slight changes to the system.

Fig 2.6 Tropical latosol soil

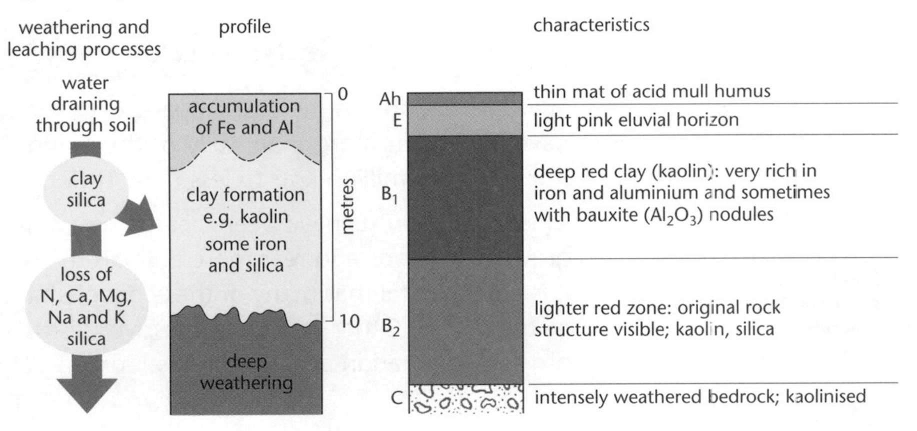

Nutrient cycling

It is important to realise these nutrients are largely held in plant tissue. What effect, therefore, deforestation?

The TRF is one of abundant biomass because:

- abundant solar energy in the tropical zone leads to rapid rates of growth
- there is plenty of water
- high daily temperatures assist rapid nutrient cycling.

Be able to deal with descriptive and interpretive data of this type.

A comparison of the nutrient stock in TRFs and temperate forests (in kg/ha)		
Elements	*Amazon lowland TRF*	*Ancient temperate deciduous forest, Cambridge*
N	2620	8320
P	24	2480
K	125	472
Ca	215	254
Mg	78	38

Figure 2.7 *The TRF nutrient cycle*

rainfall
fall out of nutrients as tissues die
BIOMASS
(10 tonnes/ha/yr)
LITTER
uptake by plants
loss in surface runoff
release of nutrients as litter decomposes (30 × higher than temperate deciduous forests)
SOIL
input of nutrients from weathered rock
loss by leaching
stores are drawn proportional to nutrients stored
transfers drawn proportional to amount of nutrient transfer

For the most part TRFs exhibit a closed system of nutrient recycling.

Figure 2.8 *Interference in the nutrient cycle by man*

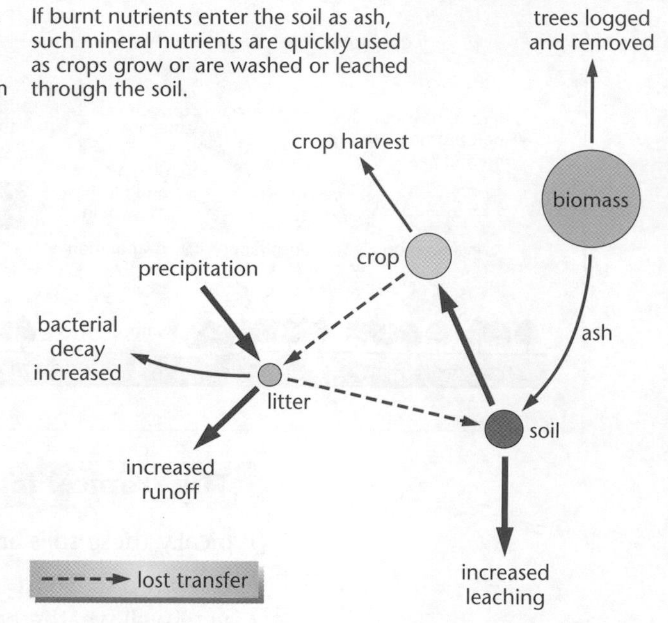

Management of the TRF

AQA A	AS
AQA B	AS
EDEXCEL A	Some U4
EDEXCEL B	U4
OCR B	U5/U6
NICCEA	U4
WJEC	Some U5

The use of the TRF is influenced by its productivity and by the distribution of its nutrient reserves. Most of these are held within the timber, the living biomass. Massive forest clearances, and even small disturbances, because of fuel ambitious development plans and the pressure of burgeoning populations cause an immediate loss of nutrient reserves along with the formation of laterites, soil erosion and leaching. Such areas therefore have to be managed if the forests are to provide for the longer term future.

The problem of destruction

At current rates all the TRF will have gone in less than 200 yrs.

TRF destruction has increased apace over the last fifty years. Since 1950 humans have destroyed more than 40% of the world TRF. Land coverage has diminished from over 18 million km^2 to less than 10.5 million km^2.

Know the problems, causes and consequences of forest loss.

Most of the forests have been cleared by traditional shifting agriculturists trying to grow more crops and rear more animals to cope with bigger families/populations. It is coincidental that many of these people happen to live in the 'poorer' countries of the world. Other causes of massive destruction include high technology dam projects, plantation and ranch development, and mining and logging operations.

Complex interrelationships: understand them!

Fifty species of plant and animal become extinct every day.

The causes of rainforest loss

KEY POINT

- pressure of population
- wood-pulp for paper
- HEP dam and reservoir construction (Brazil has plans for 31 by 2010)
- agricultural reasons:
  - plantations
  - clearance for ranching
- communication growth
- the need to 'develop'
- colonisation programmes
- mineral extraction
- economic and material resources e.g. pharmaceuticals
- timber extraction e.g. for buildings, fuel and furniture
- tourism
- over-fishing

Consequences of rainforest loss

KEY POINT

As forest clearance increases, the ecosystem becomes increasingly susceptible and fragile

effect on atmosphere and weather
- evapotranspiration
- cloud cover ↓
- insolation & re-radiation ↑
- temperature variability ↑
- air circulation changes ↑
- rainfall (related to albedo changes) ↑
- greenhouse effect (as CO_2 sequestered by trees) ↑
- seasonality ↑

effect on the water cycle
- interception ↓
- flooding – landsliding ↑
- erosion of top-soil ↑
- rainsplash ↑ — silting ↑
- leaching ↑
- pollution ↑
- fertility ↓ — productivity ↓
- water quality ↓

other consequential changes
- transmigration ↑
- migration to the city ↑
- cultural loss ↑
- loss of local habitat ↑
- debt ↑
- diseases ↑

Key
↑ = increase
↓ = decrease

Burning of the TRF sends 2 billion tonnes of CO_2/yr into the air.

Figure 2.9 Solutions and options for the TRF

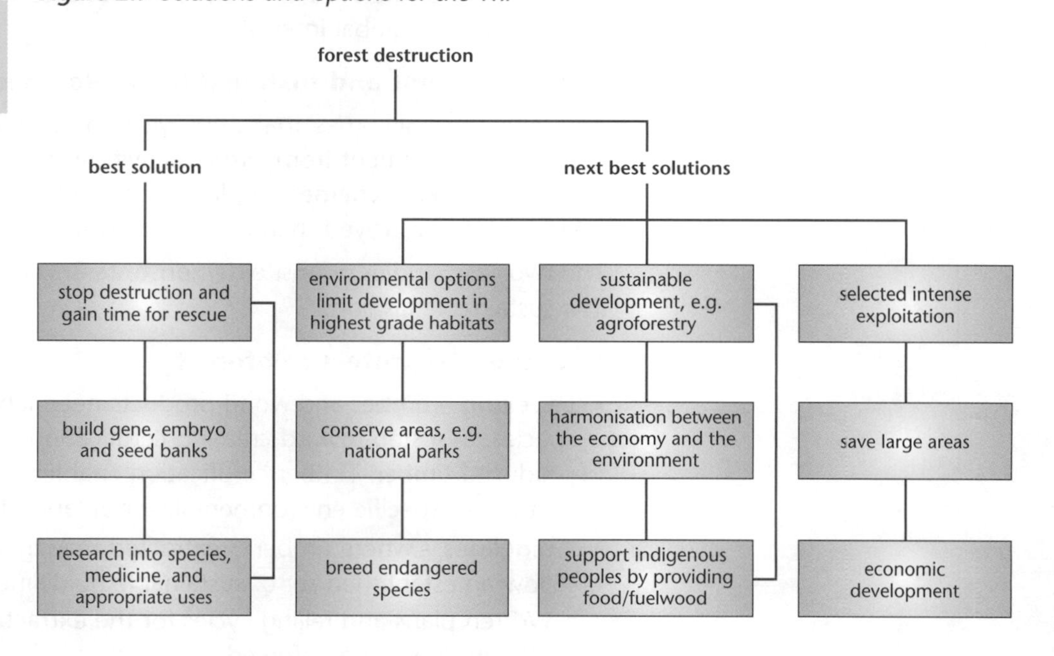

Key issues

The three most important issues related to TRFs. Know them!

The flow diagrams on the previous page raise three very important issues related to the TRF.

1 The question of deforestation and development

In the developed world, deforestation is viewed as an LEDC problem, with indigenous populations seen as either culprits or victims. Some countries fell timber commercially to generate incomes, in others the ruling minority evict the poor, forcing them further into the TRF where they fell more trees to provide building materials and fuel and clear sites for dwellings and agriculture.

A historical perspective suggests a more general relationship between deforestation and development. In the last 200 years, MEDCs felled areas of temperate woodland twice as extensive as the losses to the TRF. Perhaps another way of viewing the situation is to relate the rates of deforestation to the process of development. That is accessibility creates new roads, new markets and pioneer settlements. Could the clearing of new land be seen as development in its own right?

2 Global repercussions and eco-catastrophe

Destruction of TRF damages a rich supply of resources. TRFs are thought to contain half of the world's animal and plant species. Their genetic materials support modern agriculture (e.g. edible tree fruits, germplasm for the genetic improvement of crops), medicine (25% of our medicines have a TRF origin) and industry (e.g. perfumes, soft drinks and adhesives). If exploitation continues, considerable and untold stocks of animals and plants will disappear.

The burning of the rainforest is also known to contribute to the stock of atmospheric CO_2 but it is questionable whether they contribute any more CO_2 than the industrialised and industrialising countries of the world.

Based on the detail above we would all seem to be on the threshold of some world global catastrophe! However, it is now widely acknowledged that this is unlikely to happen. Certainly there has been violent and destructive deforestation around the globe and many species of animal and plant have been lost. However, there have been fundamental changes in how corporations and governments view TRFs. Many are now aware that poor economic development may cause inevitable violence and destruction to the TRF environment in which they live. Around the globe there is widespread support for the preservation and sensible and sustainable development of the TRFs. Politicians in countries with TRFs are generating policies that will protect global interests.

3 Management and sustainable development

Any activity or schemes that improve (e.g. thinning) or preserve (e.g. through revenue enhancement from leisure activity) can be termed **management**. Should the management schemes employed enable the TRF to replace itself at a greater rate than it is destroyed then the system is said to be operating **sustainably**.

The two case studies opposite demonstrate Brazil's resolve to manage its resources in a sustainable fashion.

Strategies to protect rainforest

- **Labelling** – timber and wood products need labelling by country of origin and species, as a step toward creating certification systems for sustainably produced timber. Timber certification enables consumers to choose a product that meets specific environmental, social, and other criteria.
- **Guidelines** – where timber is extracted management should increasingly follow an established set of sustainable guidelines:
  - Written plans and felling cycles for the extraction of wood need to be written down and enforced.

Successful extractivist forest management in Acre and Amap

Two Brazilian states have gone beyond speeches, rhetoric, and restricted "experimental" projects for the sustainable development of the rainforest.

In Acre, thanks to a long tradition of fighting for the rights of the extractivist communities, and in Amap , where sustainable development is behind all the public policies enacted by Governor Joao Alberto Capiberibe.

When he became governor in 1995, Capiberibe found a state with an R$150 million debt and a stagnant economy - but it was a state with a huge amount of forest resources. His sustainable development plans have turned the state around and, so far, are working well to bring the economy out of its rut - Amap's 7% annual growth far outshines the overall national growth. In the past few years, the state has invested approximately R$1 million on three cooperative extractivist projects, and these are succeeding to an unexpected degree.

Acre, on the other hand, has always been a state where the needs of the extractivist community were tied into the fate of the state. In Acre's recent history, especially, the plight of the rubber tappers and nut pickers has been a dire fight against the ruthless loggers and large-scale farmers intent on moving in on the tappers' land. Many people were killed in the fights that ensued, with several martyrs for the cause, the most famous being Chico Mendes, a tapper who received world renown for his fight against the destruction of the forests upon which his people depended, and was murdered for it.

People like Senator Marina Silva, who is herself of rubber-tapper descent, are fighting for the cause of sustainable development, not just in these two states, but in the whole of the Amazon as well. Governor Capiberibe notes that, with intelligent development plans, wherein raw materials are made into complete products in small-scale factories in the jungle, land becomes much more valuable and economically beneficial than if it were just farmed for soya or harvested for timber.

Source: Correio Braziliense Feb 2000

Forest management should benefit all

A project implemented in the town of Juruena, in the west of Mato Grosso, proposes to extract timber in a rational manner, with techniques that reduce the impact to the environment, and may even reduce the costs of extraction once in place. The project, implemented on a 100-hectare plot of Rohdln Industria Lignea property, has been in development for three years. The project requires that the forest be managed properly, and that workers and managers be instructed in the new methods, and requires an initial investment on the part of the company. However, the initial investment should pay off in the reduced costs of extraction. In Brazil, the process of updating forest extraction techniques to the new environmentally friendly methods is still in its infant stages, but pressure is mounting from environmentalists, and Brazil is showing a growing interest in the matter. In Mato Grosso only a small percentage of its 1500 properties under forest management systems are actually functioning in an adequate function.

Source: A Gazeta de Cuiab· April 2000

CASE STUDY

A range of strategies have been conceived to protect TRF. At A2 you would be expected to be able to elaborate on them.

- Yields/quotas need to be established and enforced.
- Felling needs to be carefully undertaken.
- Trees that are left need to be protected to enable the forest to recover.
- Silvicultural management should be undertaken to guarantee timber quality.
- Post felling, the land needs to be properly managed.

- **People power** – consumer and citizen mobilisation against global deforestation has resulted in important steps such as the mahogany moratorium in Brazil and the World Bank's forestry policy, which prohibits the Bank from supporting logging in primary tropical forests.
- **Controls** – with the support of indigenous communities, non-governmental organisations and government agencies work to change and halt destructive loans by multilateral development banks (MDBs) and to promote sustainable forest management.
- **International agreements** – e.g. the Kyoto Protocol, signed in December of 1997 by delegates from over 150 nations. Kyoto represents a vital step in efforts to address climate change by putting limits on the total of greenhouse gas emissions.
- **Debt swopping/or debt-for-nature** – this involves writing off debt if the country involves itself in conservation and sustainable development.
- **Creation of extractive reserves and protected areas** – there are 9 principal extractive reserves in Brazil at present, currently making up about 9 million acres (8%) of the Brazilian Amazon. These are large areas set aside for the

harvesting of latex, nuts, fruits, oils, and other products as a way of creating sustainable economic benefits from the rainforest without cutting it down (such use can increase annual returns to $6 680/ha compared to $1000/ha from logging). Administered by the Brazilian National Center for Sustainable Development and Traditional Peoples (CNPT), existing reserves have proved successful in halting deforestation and resolving disputes over land title. The National Council of Rubber Tappers of Brazil has set 10% of the Amazon as its target for extractive reserves.

- **Creation of biosphere reserves** – these are zoned, and protected reserve areas. There is a core area, where conservation is the key activity, and two buffer zones. In the inner zone research is undertaken. In the outer zone tourism is encouraged and resettlement of indigenous tribes undertaken. These reserves are common in the TRF areas of Thailand.

Obviously not all sustainable projects are solely focused on timber extraction, tree nurseries and replacement schemes. Increasingly projects look to establish sustainable schemes in all areas of life in the forest, i.e. craft and local item production, butterfly farming, trapping and breeding wild birds and animals and of course eco-tourism.

2.2 Tropical grassland – savanna

After studying this section you should be able to understand:

- *that savanna is typically found in areas where there is a long dry season and consistently warm temperatures*
- *how plants have adapted to the conditions prevalent in the savanna*
- *ways in which the savanna can develop*
- *how human occupants and climatic variations threaten this precariously balanced biome*
- *how poor management converts savanna to desert but careful management can halt or reverse this trend*

LEARNING SUMMARY

Distribution and characteristics

AQA A	AS
AQA B	AS
EDEXCEL A	Some U4
EDEXCEL B	U4
OCR B	U5/U6
NICCEA	U4
WJEC	Some U5

The term savanna is generally used to describe the tropical grassland biome which ranges from virtually treeless grassland to those areas with drought resistant trees or shrubs. The obvious dominant species are the perennial, tussocky and **xerophytic** (resilient to drought) grasses and sedges. Savanna seems to develop in regions where the climax community should be some form of seasonal woodland but edaphic (soil) and other disturbances prevent the climax community from forming.

Figure 2.10 *Distribution of tropical grassland – savanna*

Savanna is from the Hispanicised Amerind term for 'plain'.

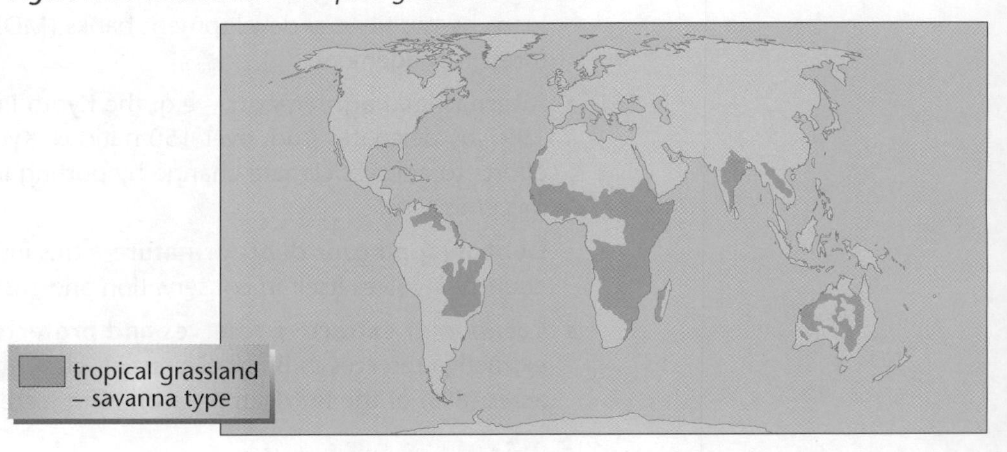

Vegetation of the savanna biome

A transect across the savanna grassland shows how the natural vegetation changes in response to the latitude and variation in rainfall (climate), see below:

Figure 2.11 A cross section through the savanna

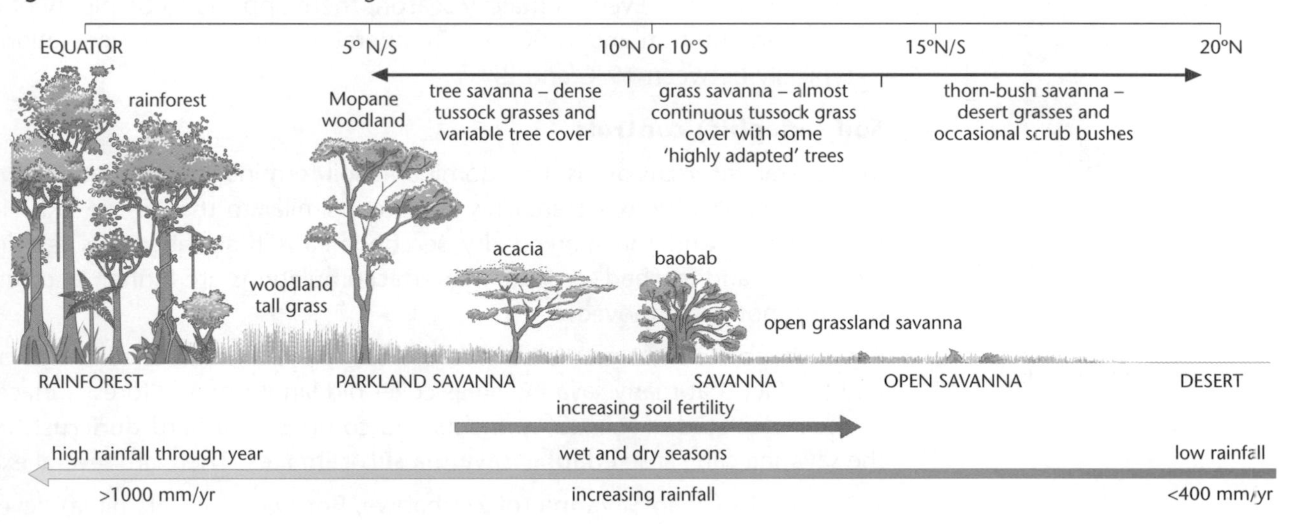

Figure 2.12 Adaptations of trees and grassland to the savanna regime

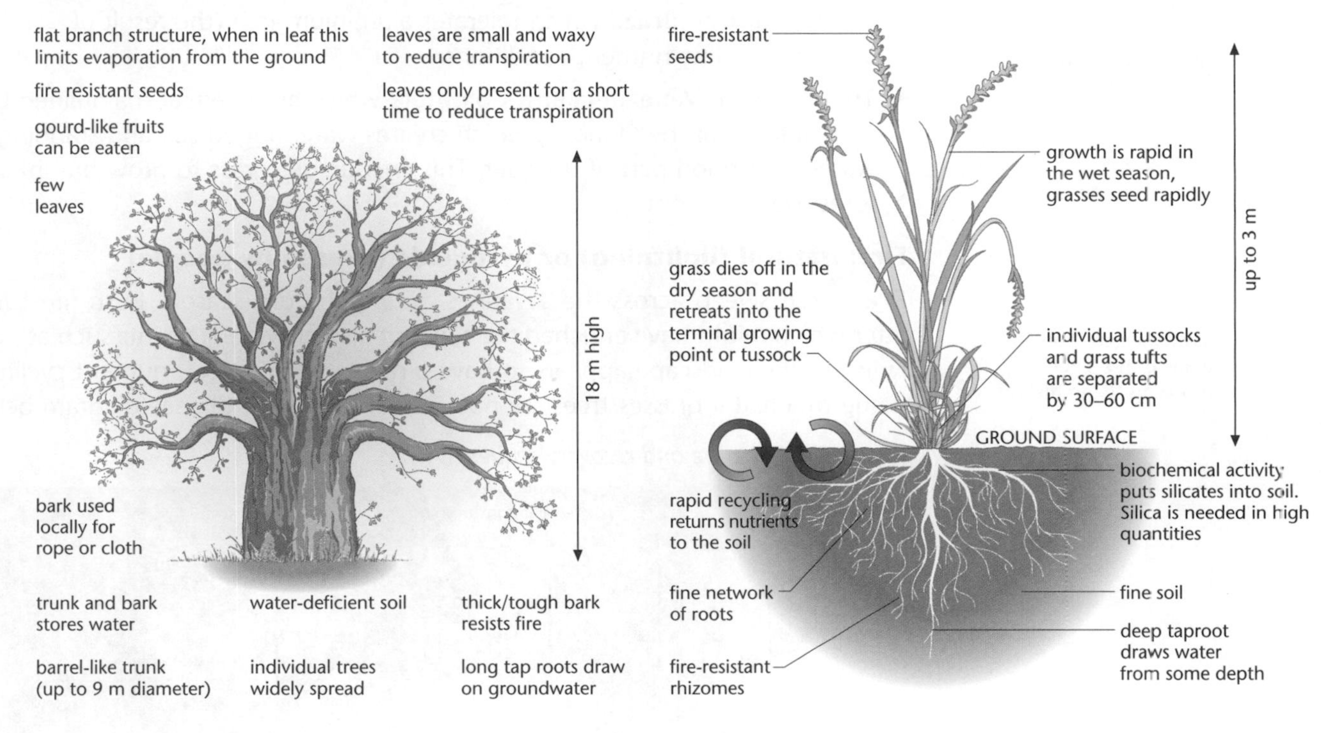

Legend says 'the devil picked up the Baobab, thrust the branches into the earth and left its roots in the air'.

Causes of the savanna biome

One would expect, given the distribution of this form of vegetation, that it might fit into a pattern based on climate alone. In reality this is not the case, the highly variable savanna biome seems to be the result of responses to a number of quite variable and separate controls.

Climate

The savanna is associated with Koppens' wet and dry climate type.

The two season savanna climate (see figure 2.13) otherwise known as **the wet and dry climate**, seems to have a dramatic effect on this grassland biome.

- The **wet season** is caused by NE and SE Trade Winds blowing in towards the equator, meeting in the low-pressure area called the doldrums (the ITCZ or

Figure 2.13
Savanna climate

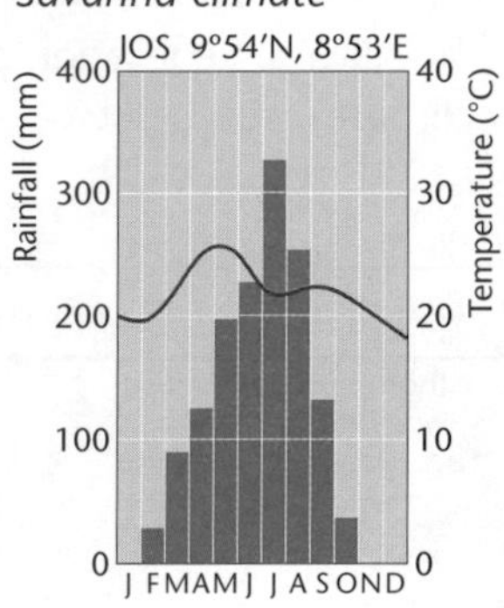

'Laterite' is the Latin for brick!

The landscape can also have a catenal effect on the soil.

Causes include: water logging of the soil, which inhibits growth; during the dry season hard baked laterites prevent root penetration.

Savanna areas suffer many of the same problems felt in the TRF.

Generally grasses are better adapted to fire than trees.

Inter Tropical Convergence Zone), which is over the Tropic of Cancer in June. This area experiences massive heating as air rises and cools, clouds form and it rains. Moisture-laden winds blowing in from the sea aggravate the situation.

- The **dry season** occurs when the ITCZ follows the sun south to the Tropic of Capricorn in December. At this time the Trade Winds blow away from the land and no rain falls. Even with a dry season, there appears to be plenty of rainfall, between 700 mm and 1200 mm. Temperatures are ideal for vegetation too, typically between 19°C and 27°C.

Soil (edaphic) controls

In general **laterization** is the dominant soil-forming process and low-fertility **oxisols** can also be expected. This is all very similar to that of the TRF. However, lower rainfall and the marked dry season means that the soil is less intensely weathered and leached, organic matter accumulations are higher, and (uniquely) silicon is not easily moved.

But perhaps the most marked control relates to the amount of soil water available and the fact that many savanna areas cover old landform and forest surfaces, most of them depleted of mineral nutrients and composed of hard **duricrust**. In effect, the savanna can cause **edaphic savanna sub-climaxes**. There are several examples.

- **The East African Savanna** (of Zimbabwe, Botswana and Namibia) developed on droughty, but nutrient-rich volcanic sandy soils, i.e. the Serengeti Plains in Tanzania (though these are largely controlled today by fire and grazing).
- The **Cerrado of Brazil** which tolerates aluminium-rich (the result of laterisation), low nutrient conditions.
- The **Llanos of Venezuela and Colombia** which are in effect maintained by the annual flood of the Orinoco, which creates waterlogged soil and standing water for a good part of the year. This encourages grass to grow but inhibits forest development.

Fire: natural (lightning) or pyrogenic (caused by people)

Fires often sweep across the savannas during the dry season. Bush fires destroy litter which might have enriched the soil, but also provide nutrients such as potash. Firing of the landscape appears to have a rejuvenating effect; nutrient cycling can come to a halt if grasses/trees are not periodically burnt off, see diagram below.

***Figure 2.14** The fire and recovery sequence*

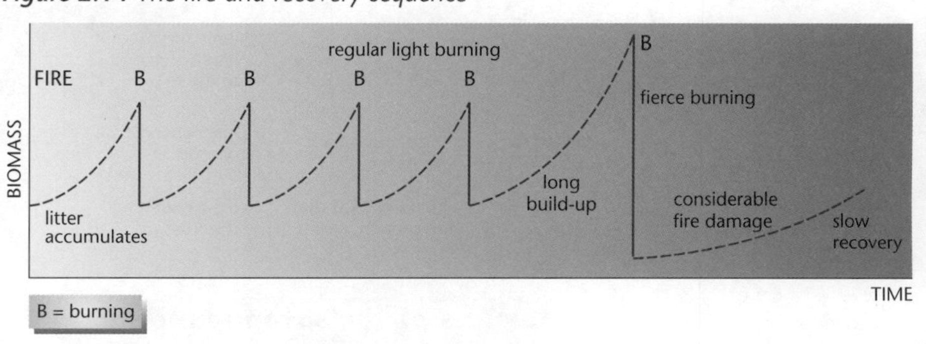

Many areas on the savanna, particularly those under pressure from population growth, are deliberately and seasonally put to the torch, to encourage the growth of grass species that will have high nutritive value for domesticated grazing animals. It is these fires that generally do the greatest damage on the savanna. The savannas of SE Asia are generally thought to have been 'man-made'. The considerable pressure on the land of Africa will ensure that fire is used as an aid to hunting and the clearing of land for some considerable time to come: e.g. in July 1998 fire destroyed all the grassland in SE Tsavo National park in Kenya; the fires had been set by nomadic herders.

Grazing

It is possible to view the savanna as a grazing sub-climax. In the natural world large mammals debark and knock over trees, grazers eat and trample seeds and inhibit/slow the growth of the larger plants. Overgrazing has contributed to both the savannas' spread and to its demise in some areas (see later section). Growing populations and the restrictions placed on nomadic tribes (like the Fulani/Tuareg) mean that grazing areas are limited and over-used. The result is that the grassland never has a chance to recover, seedlings don't grow and the grass does not develop fully. Bare ground starts to appear, evaporation increases and very rapidly a dry microhabitat develops, e.g. the Ethiopian province of Africa.

Figure 2.15 *Summary of factors affecting the savanna*

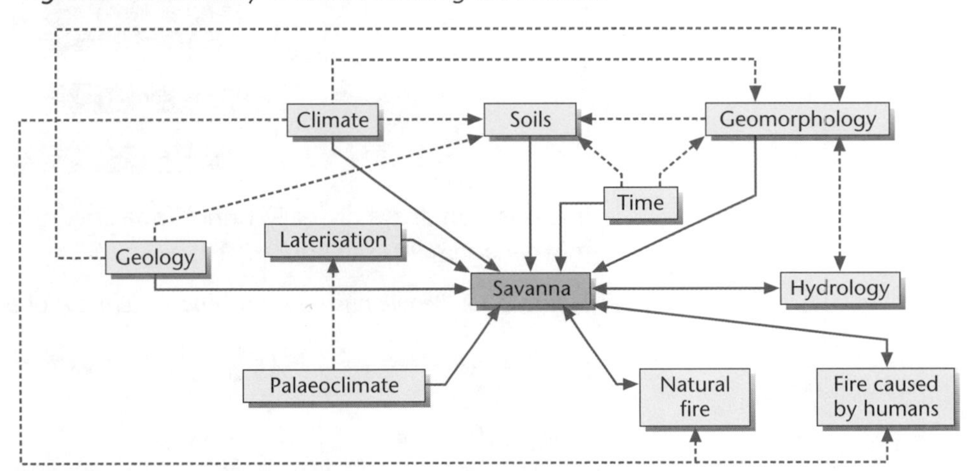

Desertification

AQA A	AS
AQA B	AS
EDEXCEL A	Some U4
EDEXCEL B	U4
OCR B	U5/U6
NICCEA	U4
WJEC	Some U5

The inevitable consequence of deforestation and savanna destruction.

Know for A2 the causes and consequences of deforestation.

Savanna areas can be very productive but are also extremely precariously placed. According to the United Nations, desertification results from various factors, including climatic variations and human activities.

There is little doubt that in recent times the world climate has fluctuated, over recent years in Africa average and above average rainfall (e.g. in Mozambique 1999) have followed long periods of drought. Over the last 30 years the precipitation falling onto the continent of Africa has been up to 48% lower in the central savanna belt. Causes of these climatic fluctuations could include El Niño, global warming (affecting the position of the ITCZ – Inter Tropical Convergence Zone) or a general global shift in weather patterns.

Figure 2.16 *WETTER DECADES*
Successive years of adequate rainfall encourage temporary settlements in the dry areas.

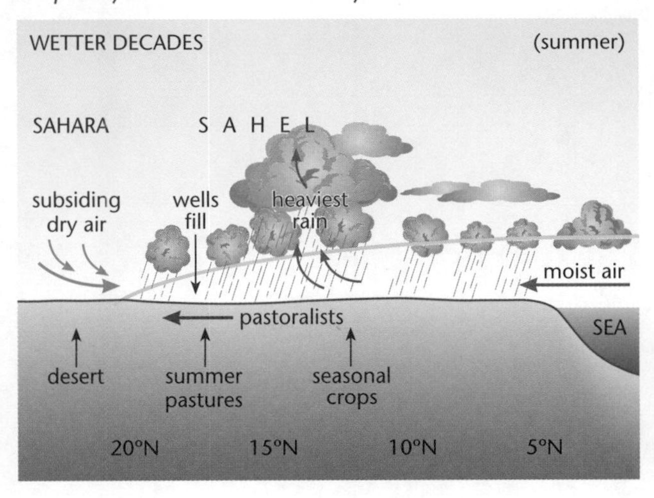

Figure 2.17 *LONG DROUGHT PERIOD*
These areas are abandoned as drought forces families and livestock to migrate.

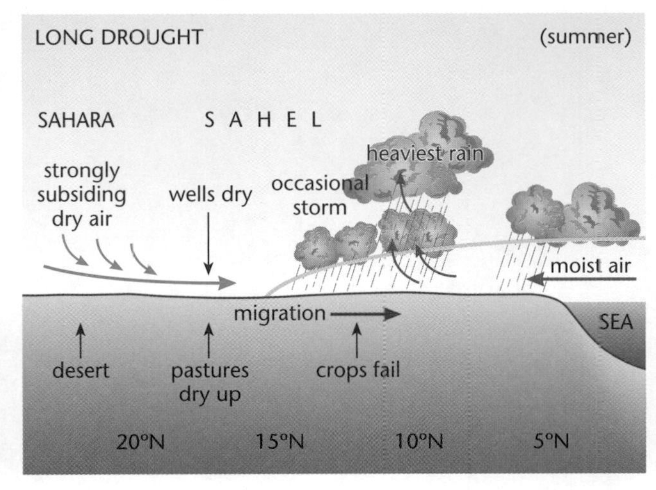

Figure 2.18 *'The savanna system out of balance'*

rains fail for several years
salinity increases
less water is stored in plants, soil and rocks
land supports fewer people and animals
more water evaporates
fewer plants grow
fewer nutrients in the soil and N.P.P drops
soil compacts
ground gets hotter
more bare ground
albedo increases
more soil is eroded by wind and water
desert conditions spread (desertification)
dust storms occur
less soil

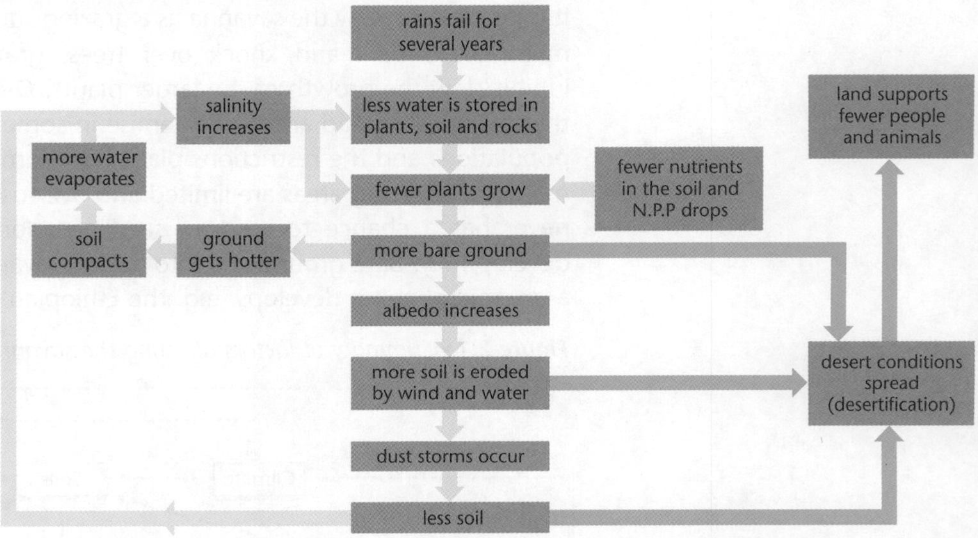

But it is humans that really bring imbalance to the savanna biome through their mismanagement, see figure 2.19:

Figure 2.19 *'People push the savanna system out of balance'*

more people and animals
more water needed
more food needed
more fuelwood is needed
famine or emergency relief
more wells drilled
people grow crops on land needed for animal grazing
more woody plants are taken from the land
land supports fewer people and animals
grazing animals are pushed on to drier areas
desert conditions spread (desertification)
overgrazing
crop land is overused
land is stripped bare
crop yields fall
land is stripped bare
salinisation increases
MORE PEOPLE + MORE ANIMALS + DROUGHT YEARS = DESERTIFICATION

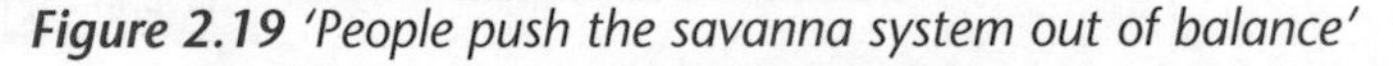

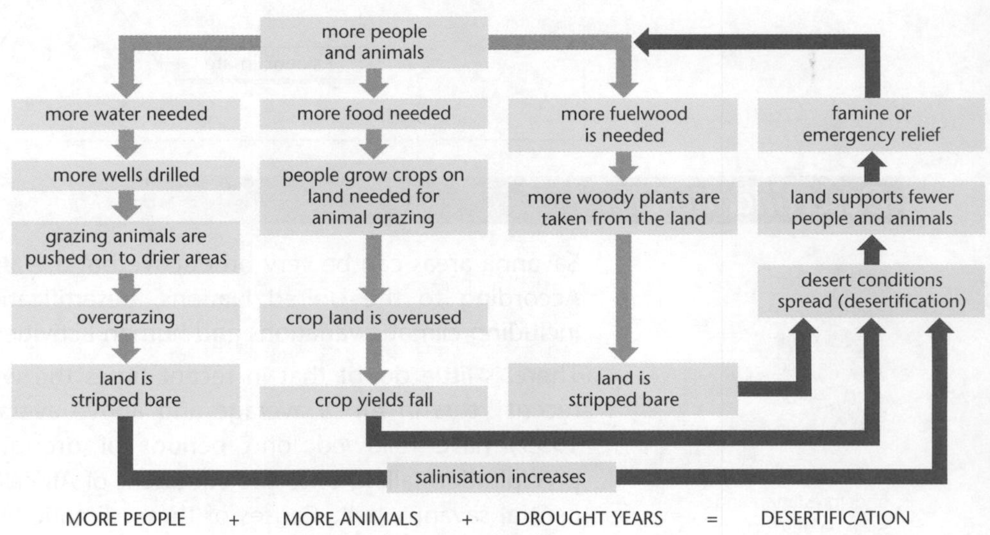

There is little doubt desertification is occurring and that its effects continue to creep insidiously forward. However, there are a range of ways in which this desertification can be managed to increase the savanna biome's resilience and recovery rates to enable it to continue providing for indigenous tribes and maintain the natural flora and fauna, see opposite:

Reduce overcultivation	Reduce overgrazing	Reduce deforestation	Improve soil conditions	Alter social and economic conditions
• Use fertilisers to increase yields • Use high yielding/drought resistant crop varieties • Use crop rotations • Use irrigation in the dry season	• Improve grazing through controlled burning of grasslands • Introduce new breeds • Improve the medical care of the stock • Rotate grazing areas • Introduce game ranching	• Use agroforestry (a combination of agriculture and forestry) • Tree planting schemes (Majjia Valley Project Niger) • Introduce alternative sources of fuels and building materials	• Use natural mulches or plastic sheets to trap moisture • Build low stone or earth walls parallel to the contours to trap sediment and surface runoff (Burkino Faso) • Plant leguminous crops to add nitrogen • Reduce the effects of salinisation • Add fertilisers and organic material	• Decrease the dependence on subsistence agriculture either by introducing commercial agriculture or developing craft agricultural based industries • Extend tourism • Provide loans and grants

CASE STUDY

The Sahel

This is the transitional zone in Africa between the Sahara Desert to the north, and the tropical forest and savanna to the south. This dry belt stretches across Africa.

The so-called environmental emergency that is the Sahel is the result of climatic variability, namely drought, and localised environmental degradation, sufficient to mean that agricultural production and livestock numbers have dropped. Talk of the Sahel today and one immediately conjures images of 50 million people overawed by economic fragility and food security problems. There has been migration from the area and some urban growth, but for the most part the region is still rural and dependent upon rain-fed agriculture and animal rearing. Three huge and long-lasting droughts struck the region late last century, killing 100 000 and wreaking havoc on the people and the land. A once prosperous and self-contained savanna people almost overnight became vulnerable.

But things are changing for the region. Reaction to the Sahelian problem has focused on:

1. famine early warning systems developed by USAID
2. soil and water conservation, through agro-forestry and other cheap sustainable methods
3. more expensive production techniques, i.e. high yielding rice, drought resistant crops, irrigation
4. repositioning and restructuring the Sahel's place in Africa. Funded by foreign aid, investment has occurred in basic industry, agricultural exports and transport.

The above has been established with full regard to the fact that population will double in the next 30 years (this problem needs to be addressed too!).

Future development of the Sahel must recognise its complex social and ecological structures. Progress has been made in the Sahel through the intervention of foreign governments. Its increasingly aware and enabled local democratic governments must ensure a productive and lasting future for the inhabitants and, importantly, the environment.

Sample question and model answer

1

Describe fully how people have intervened, altered and changed the tropical savanna biome.

Always put down the title of your chosen essay. It makes the examiner's life easy!

Concise and fairly accurate introduction. Never 'go over the top' in your introduction in these extended writing sections!

Tropical grassland or savanna is located between 5° and 15° north and south of the equator. Tropical grassland lies between the tropical rainforests and the desert areas. The climate is one of a hot dry season and a wet season. The vegetation in the savanna, because of the extremes of climate, has to be both xerophytic, to survive the long drought periods and pyrophytic to resist the frequent burning of the landscape through natural and man-made fires.

Useful to quote, but ensure you credit it to someone!

Good introduction of the terminology of the subject.

By far the biggest problem facing the savanna is the catastrophe that is desertification. This process is defined as 'the process by which arid lands, which are marginal farming and grazing areas, are transformed, by the removal of vegetation and soil into barren desert'. Clearly the environment suffers and there are estimated to be one billion people around the globe who are threatened by desertification.

A cautionary note, i.e. it is not just humans that cause desertification!

Desertification, it would seem, is the result of human, climatic and physical factors. It is claimed by many that climatic fluctuations cause this desertification. It is however unlikely this is the sole reason. Human activity and the relentless growth of population are more likely to be the key to this process, as higher populations lead to over-grazing, over-cultivation and over-use of wood for cooking and general energy requirements.

In the Sahel region in northern/central Africa natural and man-made desertification work in tandem to move the desert unstoppably southward.

To develop the human theme further, to show how man has affected the savanna of Africa:

This section constitutes the main body of the answer, paragraphed section might have been better than bullet-points?

- Removing the vegetation cover increases the albedo of the land, so there is less solar radiation to heat up the land, because the reflectivity of the land increases, less cloud forms and less rain falls. Less vegetation also means the soil gets exposed to the wind and rain, the soil gets eroded and it becomes infertile as the nutrients are gradually leached out.
- Exposed topsoil becomes hard very quickly and when the wet season comes the water finds it hard to infiltrate into the soil and flash flooding can occur. Over-cultivation also causes soils to become infertile. Prolonged drought and flooding have a long-term detrimental effect.

This is the support for your human 'argument': it is useful to use examples, these bring the essay to life.

- On average each person in the savanna uses wood every day for cooking and other energy needs. As an example a Fulani/Tuareg family would use about one hectare of wood products and debris per year. There are 50 000 000 people in the Sahel area, this means that some ten million hectares of wood is needed every year. This level of consumption is not sustainable for this environment, desertification increases.

Consider the style of what has been written so far, it is tight and informative! i.e. no waffle!

- Other human interventions lead to the savanna's mismanagement. Poor irrigation projects cause salt to rise to the surface; in the hot season, a crusty layer of salt forms on the surface actually reducing crop yields.

Sample question and model answer (continued)

Here we have the solution to the problem – good idea to include these.

If desertification is to be tackled there needs to be a sensible reduction of domesticated animals, to a level that is realistic. Further, farmers need to be encouraged to halt the spread of desertification; they can be persuaded to rotate crops rather than continue with damaging monocultural techniques that remove, endlessly, the same nutrients from the soil. Remembering that soils that lack nutrients lack vegetation and a lack of vegetation leads to desertification, in some areas farmers have literally taken the 'bull by the horns' building simple stone/earth dams and smaller stone bunds across the landscape that follow the contours. As water finds its own level, the water will tend to sink into the ground, its movement restricted by the walls and dams. Nutrients and soil are captured, the vegetation returns, crops can again be grown and desertification is slowed.

A summary-type conclusion. Same key points reiterated. Perhaps it needed a little more detail?

Man has undoubtedly accelerated desertification, by removing trees, through over-zealous tourism development, over-cultivation and mismanagement of water. The process is reversible, but it needs the co-operation of locals on a supra-national scale in Africa this means through the whole Sahel region.

Comment

This type of essay appears, on first impressions, to be conceptually easy, but such essays are difficult to answer without a great deal of detailed and factual case study material. A strategic decision seems to have been made here not to include detailed case study material, it needed more than the Fulani/Tuareg reference to bring it alive. In an examination it would have come at the bottom end of the higher 'levels'.

Practice examination questions

This resource based/decision-making question is very similar to the types that are being offered by several of the examination boards, for instance the OCR Sustainable Development Paper (Specification B). Generally these papers use pre-released resources, in some cases of up to fifteen or so pages. A far briefer resource pack is offered in this practice question! Further, you are advised to use the preceding chapter to help you out!

Rainforest Resources 1 to 3

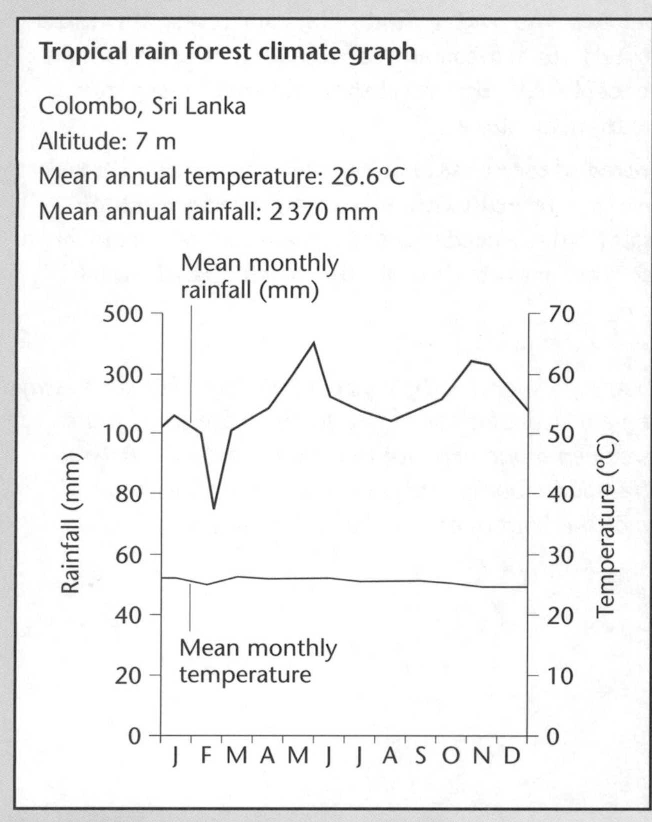

Diagram 1

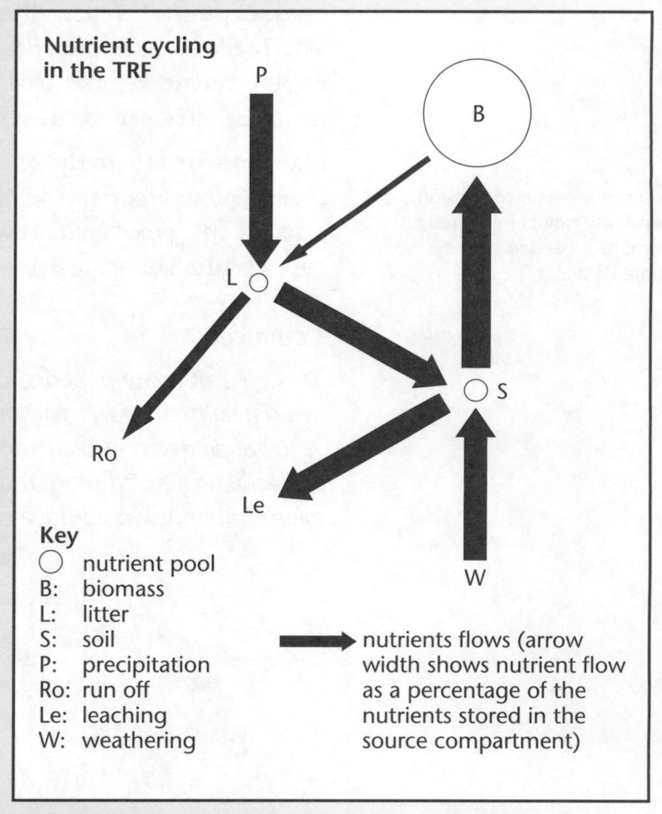

Diagram 2

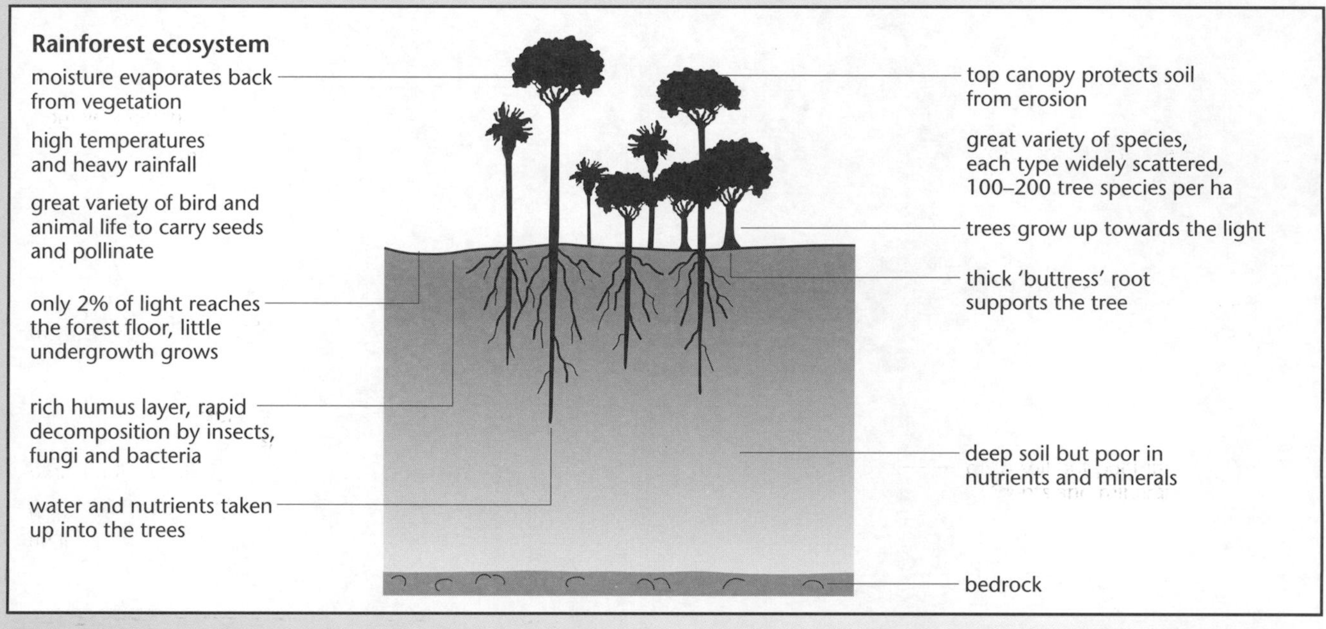

Diagram 3

AEB, AQA from 1997 Specimen papers

Practice examination questions (continued)

1 Refer to diagrams one to three.

Comment on the view that most forms of tropical development are unsustainable. (You need not just refer to forest resources. Do ensure that you look to comment on the effects on the nutrient cycle, the climate, the ecosystem and the hydrological cycle, where relevant, for whatever resources you cover.) [25]

2 Using the first part of chapter two.

Comment on the view that 'natural and modified forests provide human beings with a wealth of benefits'. [25]

3 Using the preceding chapter.

Prepare some notes for a presentation to outline the views of the governments of LEDC and MEDC countries over the fate of the tropical rainforests. [25]

Chapter 3

Arid and semi-arid environments

The following topics are covered in this chapter:

- *The causes and distribution of deserts*
- *Desert processes and landforms*
- *Desert conditions and plant adaptations*
- *Man in arid lands*

3.1 The causes and distribution of deserts

After studying this section you should be able to understand:

- *the causes of aridity*
- *the distribution of arid and semi-arid areas around the world*

LEARNING SUMMARY

Aridity

EDEXCEL B Some U5
OCR A U4

A straightforward topic but not normally examined at A2.

Aridity is a lack of water and generally it can be classified by mean annual rainfall. Temperature also has an effect as it can determine evapotranspiration. Surprisingly many areas with similar rainfall regimes to Great Britain, may be classified as arid due to other factors at work.

The 'rainfall definition' of aridity		
250 – 500	Semi-Arid	Sparse vegetation such as grassland, few trees grow
25 – 250	Arid	Plants only appear along river courses
< 25 mm/yr	Extremely Arid	Plant growth only after rainfall
These classifications cover one-third of the world		

The causes of aridity

Pressure

Sand covers 20% of the Earth's surface. Over 50% of this area is deflated desert pavements.

Deserts are found in 60+ countries of the world between 10° and 30° N and S. About one-third of the land surface of the world is classified as arid, semi-arid and/or dry.

Figure 3.1

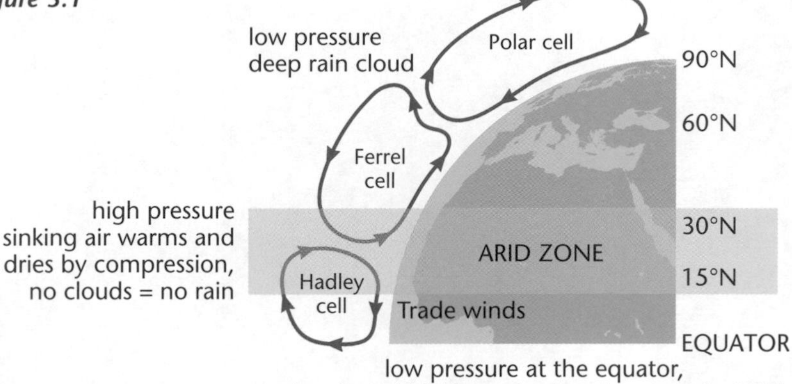

Cold ocean currents

Cold air present above such currents ensures that there is little moisture available to cool and form clouds. The coasts of Western, North and South America and Africa display such conditions. Both continents have west coast deserts just a little inland.

Key points from AS

- **Water management**
  Revise AS page 28
- **The challenge of the atmosphere**
  Revise AS chapter 3
- **Ecosystems**
  Revise AS page 82

Rainshadow and continentality

Air descending from mountainous areas warms and dries by compression, little rainfall forms and aridity is the result. Central areas of continents are dry because air moving over landmasses does not absorb large amounts of water vapour. During the last ice age, conversion of water to ice resulted in larger continental areas. This extreme continentality is thought to have facilitated the spread of deserts during the ice age.

World wide distribution of deserts

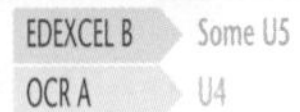

The majority of the world's most arid areas lie between 15° and 30° North or South of the equator (see map).

Figure 3.2 *The world's major deserts and associated ocean currents*

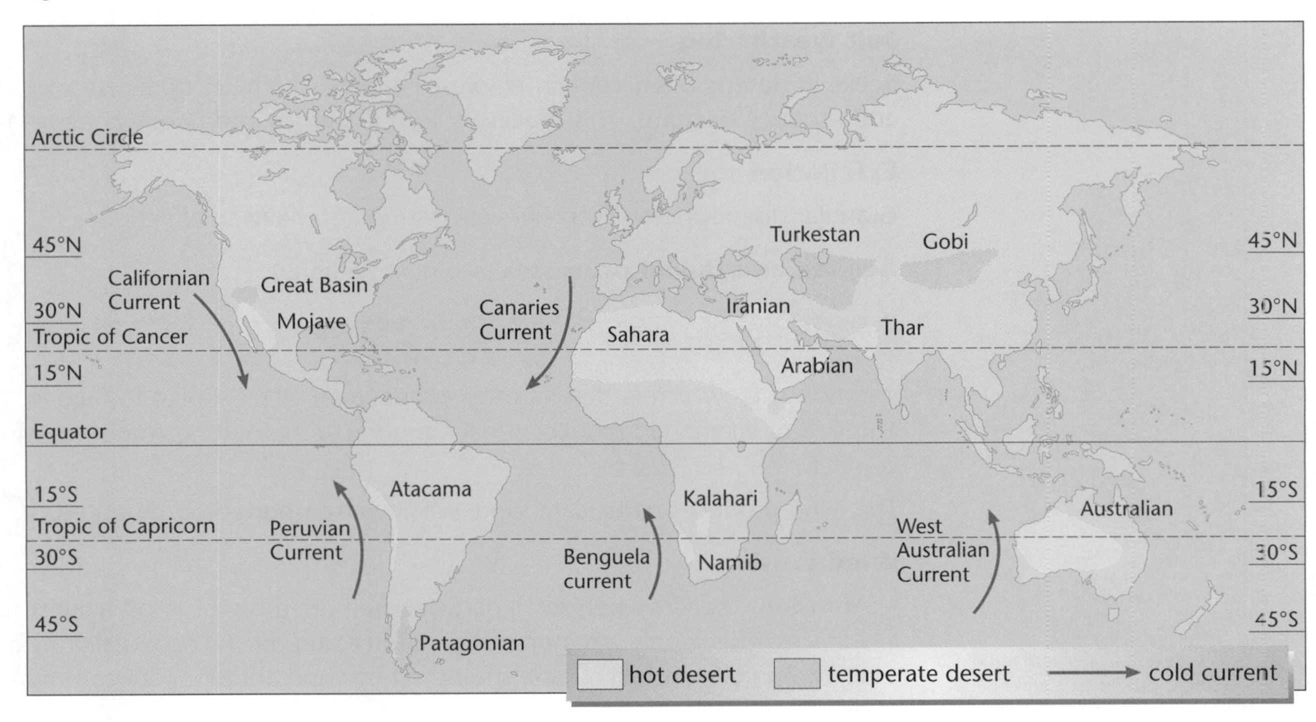

3.2 Desert processes and landforms

After studying this section you should be able to understand:

- *the processes of weathering in deserts*
- *that water and wind processes cause differing landforms*
- *that the origin of some landforms is difficult to ascertain*

LEARNING SUMMARY

With regard to desert landforms, controversy has raged for many years over the part played by water and wind (**aeolian effects**) in forming desert landform features.

Processes

Desert surfaces

A range of desert surfaces exist:

- **Ergs** – sandy deserts/sand seas, common only in about 30% of deserts. Their distribution seems to be climate linked (i.e. less than 150 mm of rain).

- **Pavements, gibber plains or reg** – form as a result of wetting and drying. They are hard, rock covered areas.

Weathering

Desert weathering is a controversial topic.

Chemical weathering is limited by:

- the lack of water
- low rates of penetration into the rocks
- the amount of capillary action
- the alkaline nature of the chemicals taken into rocks creating no aggressive acids.

Other weathering processes include:

Salt weathering

Rocks in deserts often contain efflorescent salts which set up stresses in the rock and produce fractures. This process is seen in porous and poorly cohesive rocks.

Exfoliation

Granular disintegration and chemical rotting also have an effect.

Aeolian and fluvial processes (see next sections)

Aeolian processes

Winds that blow across deserts often produce an effect similar to fluid in motion. The lack of vegetation reduces surface roughness permitting smoother wind/land contact.

The wind produces particulate sand, which is transported or deposited.

Wind erosion

In deserts water and wind processes are equally important. Know how they relate together!

- **Abrasion** – occurs when small particles are hurled by the wind against rock surfaces. This is only a minor erosional force and mostly occurs slightly above the ground. **Ventifacts**, rocks smoothed by wind abrasion, are common in deserts.
- **Deflation** – wind blows away rock waste and lowers the desert.
- **Attrition** – rock particles rub against each other and wear away.

Aeolian transport

- **Saltation** moves small particles in the direction of the wind in a series of short hops and skips. It normally lifts sand-size particles no more than one centimeter above the ground, and proceeds at one-half to one-third the speed of the wind. A saltating grain may hit other grains that jump up to continue the saltation. The grain may also hit larger grains that are too heavy to hop, but that slowly creep forward as they are pushed by saltating grains. This is called **surface creep**. Velocity is of course an important variable, a critical velocity has to be reached before particles will move (see below).

Figure 3.3 *Saltation*

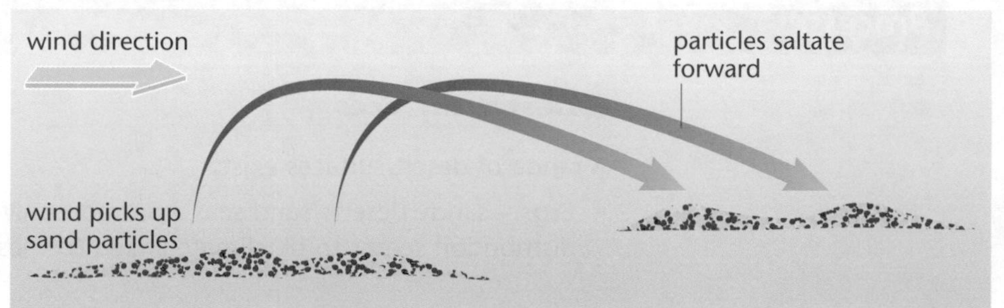

Deposition

Three processes have been recognised:

- **sedimentation** – settling occurs and there is no further effect on other sand particles
- **accretion** – occurs when sand grains come to a rest
- **encroachment** – the process of continued growth of sand accumulations. Once sand has accumulated it traps more and more sand, ripples turn into dunes, and dunes into '*draa*'.

Fluvial processes

Desert lakes are generally ephemeral and are called playas. They vary in size from a few metres to several thousand km^2. They are very salty.

Three main types of river are found in desert areas:

- **exogenous rivers** – sources outside the desert
- **endoreic rivers** – these form near the desert and never show beyond it
- **ephemeral rivers** – these flow for only part of the year.

Drainage systems

The mountain areas of deserts and the lowland deserts have hugely different drainage systems.

KEY POINT

Mountainous areas

A high amount of scouring occurs producing very rocky beds and lots of debris/sediments in the upper areas of mountains. As slope decreases and sediments concentrate in lower areas so there is a rise in the deposition of alluvial fans at changes in slope.

Lowland areas

The nature of the surface over which water flows determines the drainage pattern. Few permanent rivers exist; where they do, they are shallow, sandy, straight and lack sinuosity. In flood conditions they are choked with sediment.

Landforms

EDEXCEL B — Some U5
OCR A — U4

Features produced by wind erosion

Rock pedestals

Wind sculpts stratified rock into pedestals by wind abrasion and weathering, e.g. Gava Mountains, Saudi Arabia (see figure 3.4).

Yardangs (width to depth of 4 : 1)

The Sphinx at Giza may be a modified yardang!

A ridge and furrow landscape. Wind abrasion concentrates on weak strata; leaving harder material upstanding (see figure 3.5).

Zeugen

Wind abrasion turns the desert surface into a ridge and furrow landscape, e.g. various areas in Bahrain (see figure 3.6).

Figure 3.4 Rock pedestal

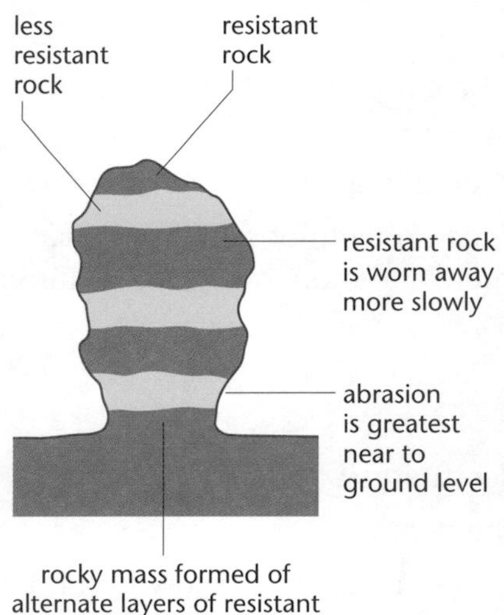

Figure 3.5 Yardangs

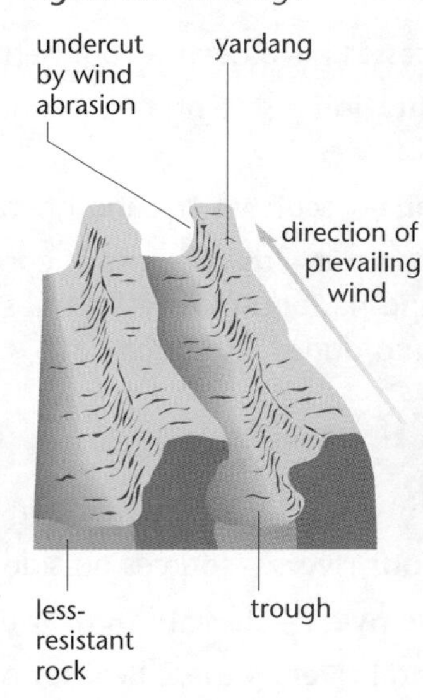

Figure 3.6 Zeugen

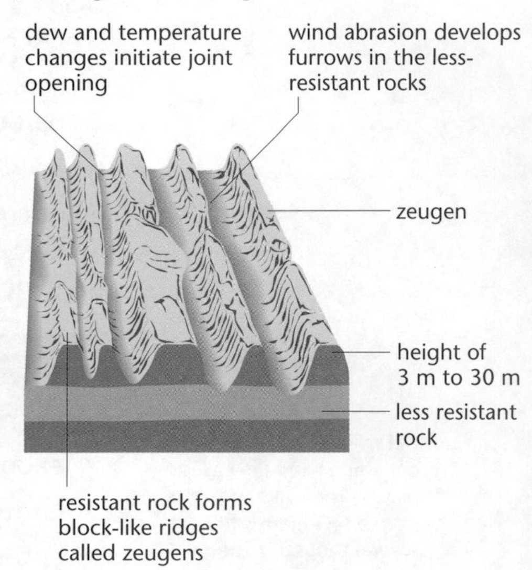

Inselbergs

Wind (and water) attacks the original surface leaving round-topped inselbergs (through **exhumation**). The material removed has a deep-seated 'decay' origin and may display extensive 'unloading' (subsurface weathering). There are two major forms: domed inselbergs (**bornhardts**) and boulder inselbergs (**Kopjes, rubbins**), e.g. Matopos, Zimbabwe (see AS Study Guide page 69).

Good diagrams and supportive labelling is a must if you hope to convey landform understanding.

Figure 3.7 Bornhardt formation

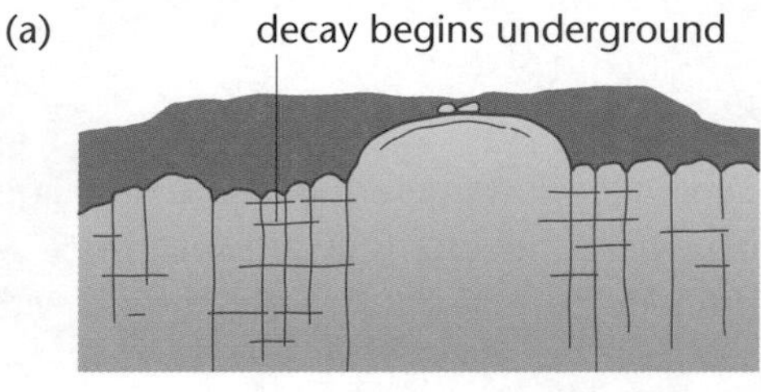

(b) weathering continues as the inselberg breaks surface

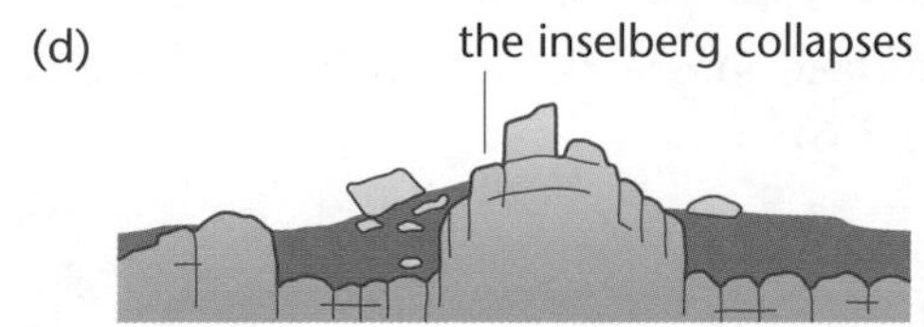

Figure 3.8 Kopje formation

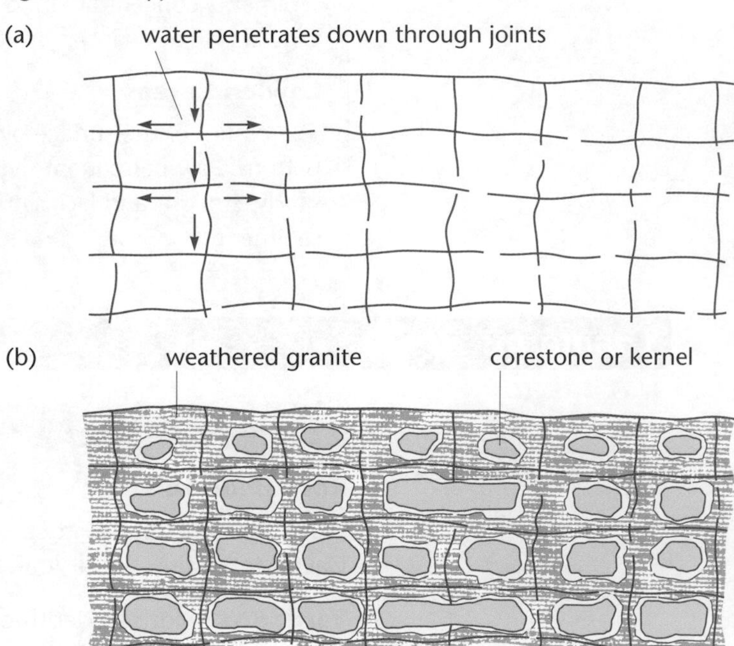

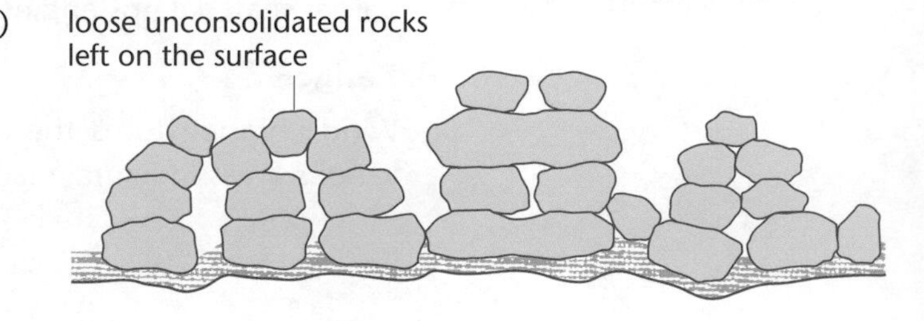

Deflation hollows

These are caused by the removal of fine particles by the wind, lowering the surface and creating a hollow, the best known example is the Qattara Depression.

Figure 3.9

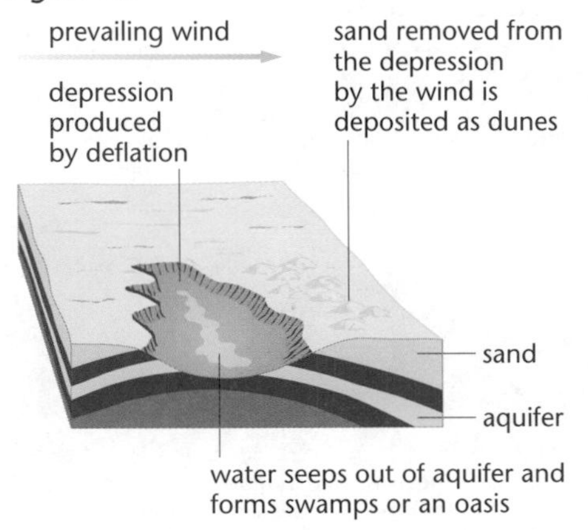

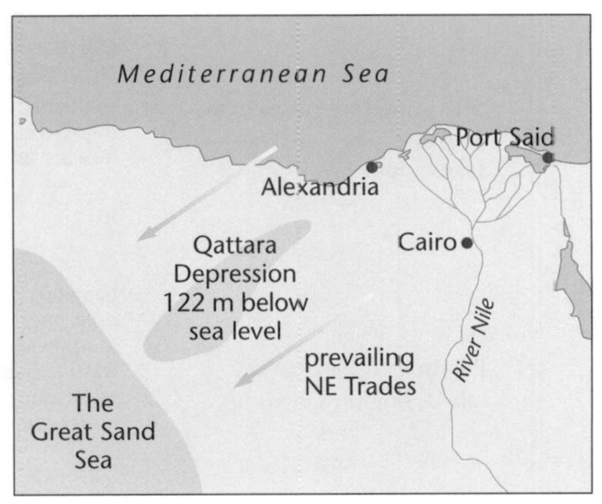

Features produced by wind deposition

Wind-deposited materials occur as sand sheets, ripples and dunes.

Sand sheets

These are flat areas of sand with sand grains that are too large to saltate. 45% of depositional surfaces are of this type, e.g. Selima in South Egypt.

Dunes

The wind eventually blows sand into a network of troughs, crests and ripples that are perpendicular to the wind direction. They are the consequence of saltation (see below).

Figure 3.10

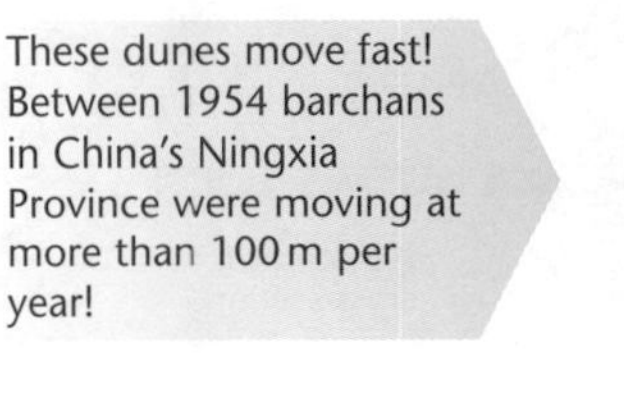

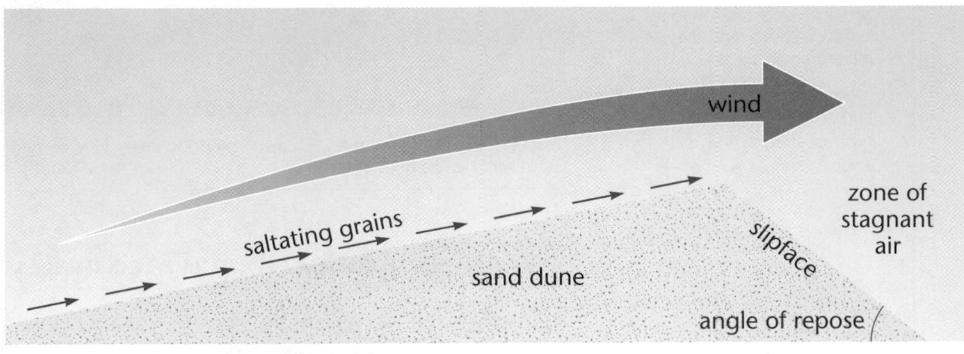

Accumulations of sand build into mounds and ridges, they become a dune when the slip face is about 30 cm high. Dunes grow as sand particles move up the gentle upwind slope by saltation and creep. They fall onto the slipface inducing movement.

> **KEY POINT**
>
> **Sandstorms**
>
> Sandstorms are a seasonal hazard in North East Africa, and are called **Khamasin** (fifty) for the number of days on which they occur. They strike around April with the onset of warm conditions. Hot rising air lifts dust up to 4500 metres above the desert. Returning as brown rain in winds of up to 110 kph it closes airports and causes many accidents. Annually at this time up to 20 people die due to the sandstorms. When the sand moves west it can destroy coral reefs in the Caribbean and has been linked to hurricane formation.

Outside of the USA, sand dunes make up 25% of arid landforms.

Five basic dune shapes have been recognised: **crescentic**, **linear**, **star**, **domes** and **parabolic**. Ralph Bagnold, an engineer, working in Egypt, prior to the Second World War recognised the main dune types, the **crescent or barchan dune** and the **linear or seif dune systems** (seif is Arabic for sword).

Star dunes are pyramidal in shape. They are the tallest dune.

Parabolic dunes have 'arms' pointing upwind.

Figure 3.11 *Barchan dunes*

(Barchan means 'active dune'. These dunes move!) Found in Northern Chad,Taklamakan Desert in China, Namib Desert in Namibia.

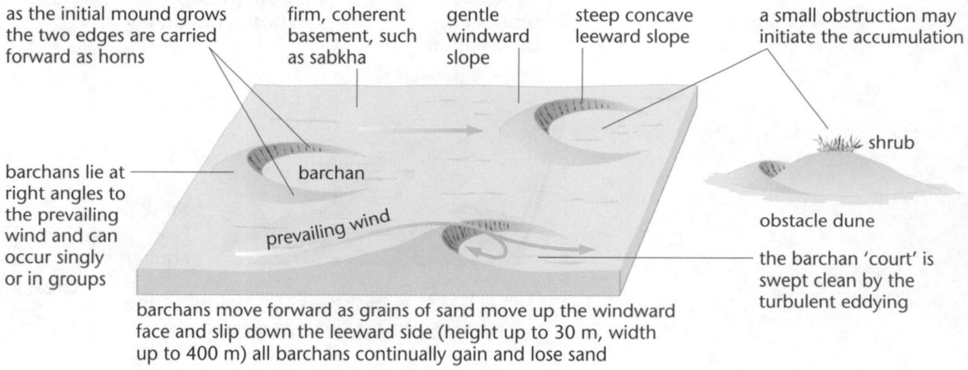

as the initial mound grows the two edges are carried forward as horns
firm, coherent basement, such as sabkha
gentle windward slope
steep concave leeward slope
a small obstruction may initiate the accumulation
shrub
barchans lie at right angles to the prevailing wind and can occur singly or in groups
barchan
prevailing wind
obstacle dune
the barchan 'court' is swept clean by the turbulent eddying
barchans move forward as grains of sand move up the windward face and slip down the leeward side (height up to 30 m, width up to 400 m) all barchans continually gain and lose sand

Figure 3.12 *Linear dunes (seif, transverse dunes or draa)*

As to formation these dunes are either:
- the result of obstacles getting in the way
- an erosional phenomenon
- as products of a vegetated landscape or,
- as products of complex wind regimes, secondary wind flow patterns (and large amounts of sand)

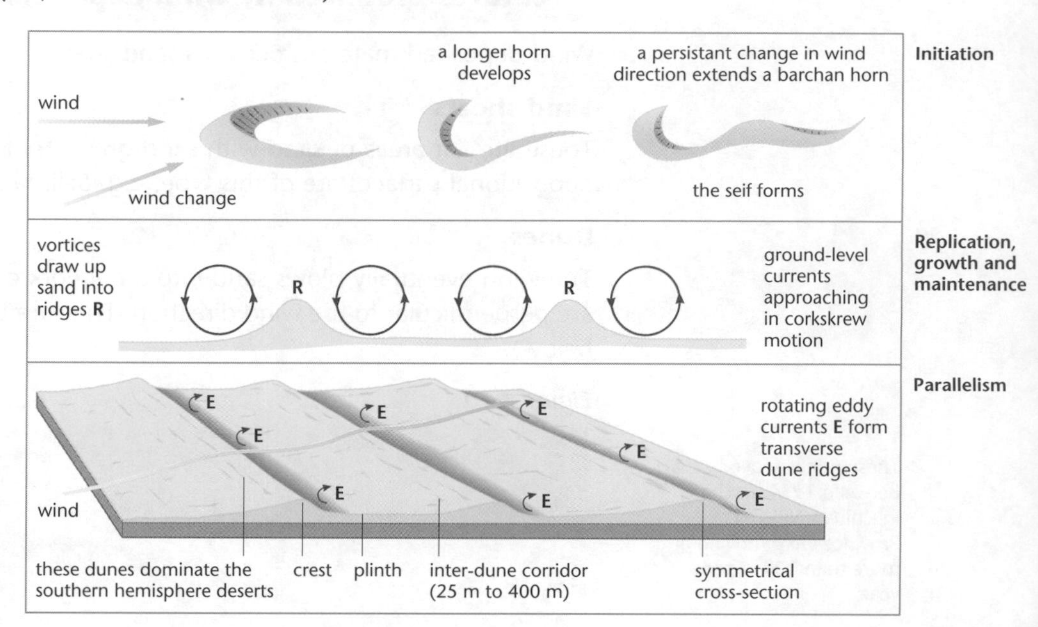

seifs have a height of 100 m and are up to 190 km long! length > width, and they are regularly spaced

You should draw this type of diagram with your essays at A2 Level.

Initiation
wind
a longer horn develops
a persistant change in wind direction extends a barchan horn
wind change
the seif forms
Replication, growth and maintenance
vortices draw up sand into ridges R
R
R
ground-level currents approaching in corkskrew motion
Parallelism
E
rotating eddy currents E form transverse dune ridges
wind
these dunes dominate the southern hemisphere deserts
crest
plinth
inter-dune corridor (25 m to 400 m)
symmetrical cross-section

Features produced by water in deserts

Rain does fall occasionally in deserts, and desert storms are often violent. A record 44 mm of rain once fell within 3 hours in the Sahara. Large Saharan storms may deliver up to 1 mm/minute.

Normally dry stream channels, called **arroyos** or **wadis**, can quickly fill after rain, and **flash floods** make these channels dangerous.

However, the evolution of arid landforms is often affected by events that occurred long ago. Past climatological conditions, reflected in many desert landforms, began to develop during pluvial periods several thousand years ago.

More people drown in deserts than die of thirst!

Channels which were once part of a perennial drainage system now receive the run-off from torrential storms, unhindered by vegetation, leaving deep alluvial debris over the wadi floor, though wind deflation may later remove much fine surface material. Surface channels on old shield areas often lead into inland drainage depressions. Many of these lake depressions have been filled by materials washed forward from the foot of the enclosing mountains, where streams entering the basin have deposited their load and created alluvial fans. In some the fans have coalesced to form a mass of material known as a **bajada**. Some of the material from the bajada washes forward over the pediment levelling the landscape with unconsolidated deposits (see figure 3.13).

Water can also dissect out deep wadi systems. Stream-dissected scarps overlook the older plains, where detached **outliers** form **mesas, buttes** and **pinnacles**.

In the USA 32% of arid landforms relate to the action of water in deserts.

Figure 3.13

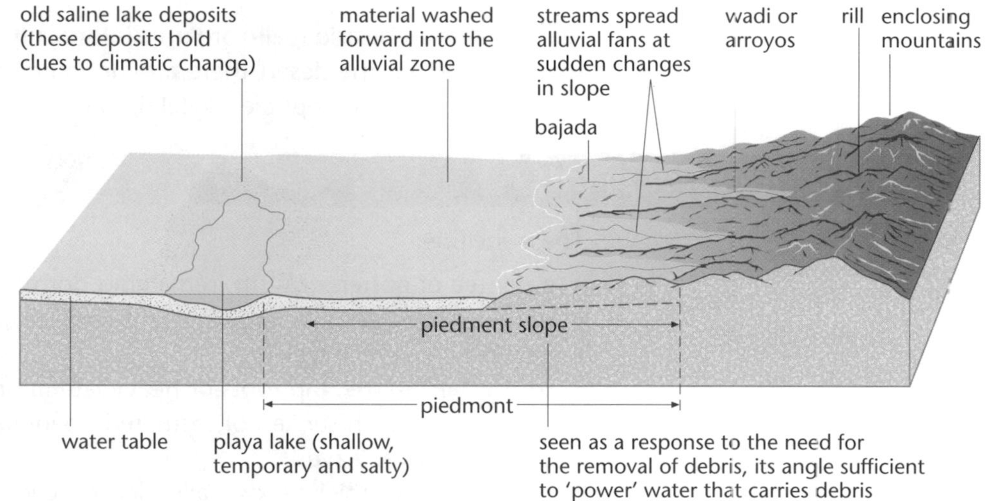

- **Sabkhas** – an occasionally flooded desert area with an extensive thickness of evaporates (salts), usually coastal in nature.
- **Duricrust** – hardened areas of minerals that cover deserts. They are impermeable and thought to be due to weathering.
- **Desert varnish** – iron and magnesium oxides and silica coat rocks. Due to evaporation.

Figure 3.14

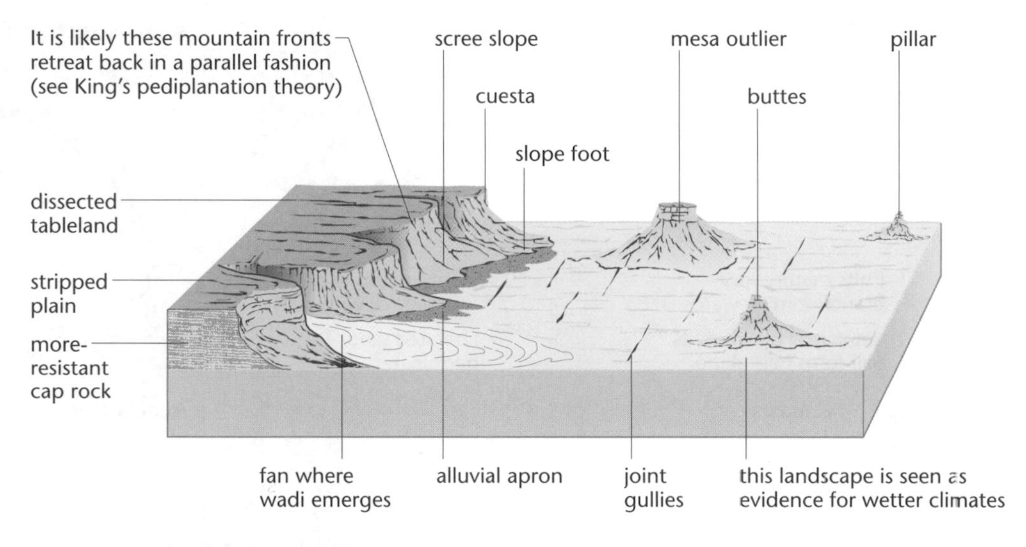

The highest salt water lake in the world is in the Qaidam Depression, China.

Ground speed records are commonly established on Bonneville speedway. The space shuttle lands on Rogers Lake Playa, at Edwards Air Force Base, California.

CASE STUDY

Lake Bonneville – a playa within a desert

Lake Bonneville is the relic of a large lake that existed during the last ice age (12 000 years ago), though it probably has an origin from 50 000 years ago. It was some 52 000 km^2 almost 300 m deep and was 1525 m above sea level. Climatic change caused the lake to fall below the lowest outlet. When the water evaporated it left an arid depression composed of billions of tonnes of salt and other minerals. The remnants of Lake Bonneville include Utah's Great Salt Lake, Utah Lake and Sevier Lake.

Equifinality

Few landforms have proven origins. Many landforms can come about through different processes and quite different conditions, e.g. pediments, dune formation inselbergs and deflation hollows may have many origins.

3.3 Desert conditions and plant adaptations

LEARNING SUMMARY

After studying this section you should be able to understand that:

- *plants have to adapt to survive in the desert environment*
- *soils have a unique appearance and composition in desert areas*

Many plants in the arid realm are **physiologically specialised**, adapted in form and structure. Within the desert there are innumerable niche locations for plants even though the soil is susceptible, skeletal, saline and immature.

Adaptations of plants

EDEXCEL B Some U5
OCR A U4

These include:

- a degree of **ephemeralism**, remaining dormant in the soil as fruits or seeds
- unique **dispersal systems**, i.e. barbs and bristles
- **xerophytic**, water-seeking
- **root adaptations**, tap roots or heavy lateral branching
- small leaves, with sunken or restricted openings
- pale, reflective, leaves
- hairs, spines or thick waxy–walled leaves, evolved to replace what might nominally exist

Figure 3.15

The Rose of Jericho

0 5 10 cm

1 fruits mature, dry branches roll inwards
2 detatched by wind, its ball-form rolls along
3 plant uncurls in moist location (or with rain)
4 liberated seeds germinate

A cross-section through xerophytic desert grass

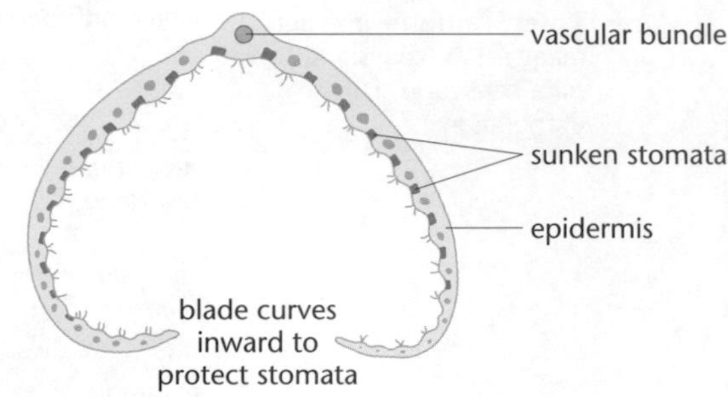

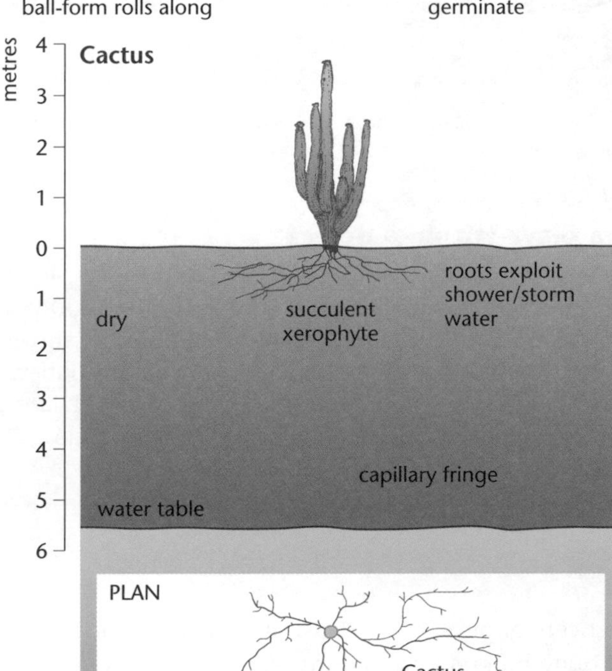

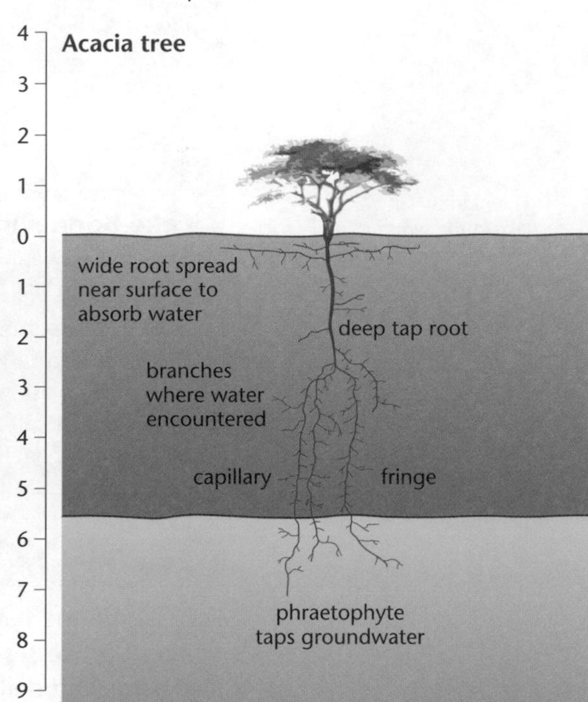

The lack of vegetation in deserts means that adaptations are necessary. A favourite topic for examiners to pick on!

- **succulents**, plants capable of storing vast quantities of water
- plants with a small surface-to-volume ratio
- **cell sap variations** allow varying amounts of water to escape
- ability to tolerate desiccation
- **halophytic**, salt tolerant, e.g. Creosote plant.

Deserts typically have a plant cover that is sparse but enormously diverse. The Sonoran Desert of the American Southwest has the most complex desert vegetation on Earth. The giant saguaro cacti provide nests for desert birds and serve as the trees of the desert. Saguaros grow slowly but may live 200 years. When 9 years old, they are about 15 cm high. After about 75 years, the cacti are tall and develop their first branches. When fully grown, saguaros are 15 m tall and weigh as much as 10 tonnes. They dot the Sonoran and reinforce the general impression of deserts as cacti-rich land.

Although cacti are often thought of as characteristic desert plants, other types of plants have adapted well to the arid environment. They include the pea family and sunflower family.

Soils

Soils that form in arid climates are predominantly mineral soils with low organic content. The repeated accumulation of water in some soils causes distinct salt layers to form. Calcium carbonate precipitated from solution may cement sand and gravel into hard layers called '**calcrete**' that form layers up to 50 m thick.

Caliche is a reddish-brown to white layer found in many desert soils. Caliche commonly occurs as nodules or as coatings on mineral grains formed by the complicated interaction between water and carbon dioxide released by plant roots or by decaying organic material.

3.4 Man in arid lands

After studying this section you should be able to understand:

- *that deserts are remarkable areas, and if managed correctly can be very productive*
- *the importance, and effects, of irrigation in the desert realm*

LEARNING SUMMARY

Settlement patterns

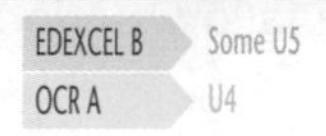

The effects of people on the arid and semi-arid lands may be seen at three overlapping cultural/technological levels.

- Small groups of people, like the bushmen of the Kalahari and the Australian aborigines, live a semi-nomadic food-gathering and hunting life, and have adapted to their environment with remarkable efficiency. Over the years, the hunter-gatherers have had little lasting impact on their environment.
- By contrast, pastoral nomads seasonally cultivate selected areas. Pastoral nomads and shifting cultivators have greatly affected the natural flora and fauna. They respond to physical conditions through their mobility.
- Many millions live a settled life in a relatively moist environment within arid zones near oases, on a flood plain or delta, or in an area irrigated by water brought from afar. Irrigation, with its associated land-use and settlement, has changed entire ecosystems, and ecological repercussions have been felt far beyond the irrigated areas.

Irrigation: a chronology

EDEXCEL B | Some U5
OCR A | U4

The growth of population means that even marginal land has to be settled and used.

Irrigation is one way man has been enabled in this 'zone'. Know others too.

Benefits and problems caused by irrigation

Benefits

Traditionally the people who settled at oases used simple technology to raise and distribute water locally through aqueducts and underground channels or **qanats**. Evaporation losses from this system are minimal, though some water is lost by infiltration.

Other methods were to use shallow, gravity-fed channels, blocked by mud packing as required; cisterns; underground caverns; depressions dug to catch run-off and retain water moving by gravity through the sub-soil; and the creation of lakes during the rains by blocking stream beds with boulders and earth dams.

Today electric or diesel pumps raise groundwater from greater depths; massive dams create deep lakes stretching back hundreds of kilometres. Metal pipelines of decreasing diameter carry water from reservoir to the field, and release it through nozzles, or tiny drip feeds, to the plants.

Problems

Problems such as water logging and saline accumulation. Water not used by crops, lost by evapotranspiration, or drained away, accumulates as rising groundwater. In desert conditions soils rapidly acquire salt from the evaporation of dilute saline irrigation water. Water acquires more salt as it slowly moves up through the ground. The land eventually becomes too saline for crops to tolerate.

Pakistan: an 'irrigated' country

In Pakistan, the whole hydrological system of a huge river, the Indus, has been transformed, as engineering systems have evolved progressively to control and distribute its waters.

Most of the flow in the Indus system is from catchments in the Himalayas and its foothills. Monsoon breaks in June and rivers reach peak flood-levels in the foothills in July–August, causing a flood wave to pass downriver. From September the river levels fall, then snowmelt brings another rise in March.

Irrigation has brought some problems for Pakistan.

- Dams and diversions, have caused floodplains to be deprived of alluvium and its nutrients, and to need expensive fertiliser.
- There is increased danger from infections like bilharzia.
- Where year-round cultivation replaces cropping with a dry fallow period, crop pests may thrive on perennial food sources.
- There are also many instances of plant pests and diseases being carried along water channels.

Despite these problems through the 19th and 20th centuries, irrigation projects have continued apace. Inundation canals were built to carry water from cuts in the bank to land parallel with the river. Gated barrages were constructed to raise the water level and allow accumulations of alluvium to pass downstream. Water could now be diverted throughout the year. Large canals were built to transfer water from the western rivers to supplement those further east. More recently, high dams in the mountains have provided storage and hydro-electricity. Perennial canals now serve most of the cropped land. Despite the aridity, over half Pakistan's workforce is engaged in agriculture.

Sample question and model answer

1

The sketch below illustrates a type of dune found in hot deserts.

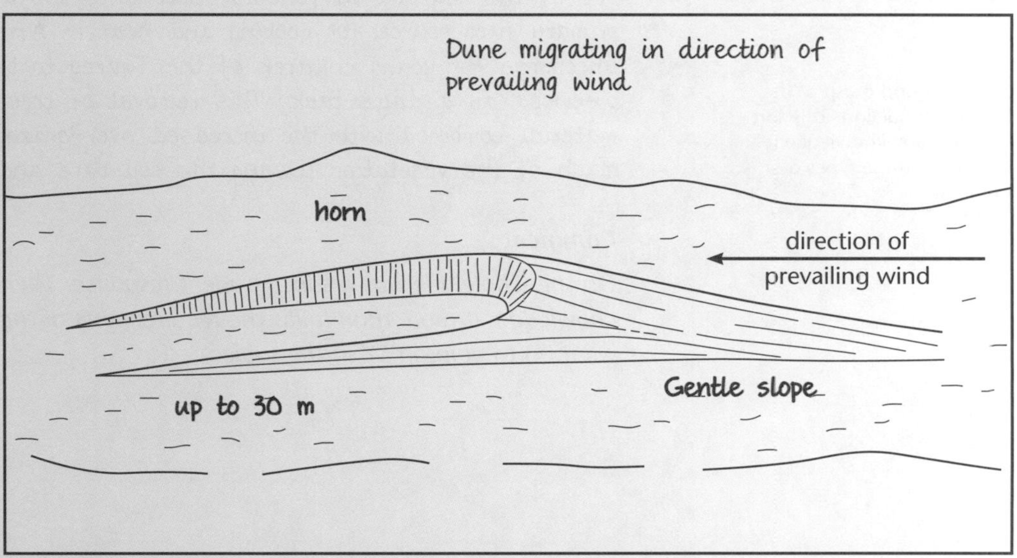

Structured landform questions invariably involve label completion.

Dune migrating in direction of prevailing wind

horn

direction of prevailing wind

Gentle slope

up to 30 m

Diagram 1

In most A2 structured questions you might expect to start with 'definitions'.

(a) (i) Identify the dune type and add three additional labels. [4]

Barchan dune

(ii) Name one other type of sand dune. [1]

Alternatives include seif, star and dome dunes.

Parabolic dune

(b) Name one landform produced by the erosive effects of wind. [1]

Alternatives include zeugens and rock pedestals.

Yardangs

(c) Suggest three physical reasons why deflation is such an effective process in arid and semi-arid environments. [3]

It would not be good enough just to state the obvious. It is important to develop ideas.

Lack of vegetation in arid/semi-arid regions means the soil is not bound in by roots, there is no protection – thus it is very unstable and easily removed by wind.

The absence of water in hot arid/semi-arid areas means that there is nothing to bind the particles together and weigh them down making deflation more effective.

Is overgrazing a physical process? Better to talk of a lack of fluvial infill, and salt weathering aiding deflation.

Human activity in the semi-arid 'Sahel' has led to 'overgrazing', with population growth leading to increasing numbers living and raising their livestock on the desert periphery. Over-grazing and destruction has damaged the infrastructure of the soil loosening it by trampling and removing vegetation.

(d) With the aid of examples, suggest two reasons for the continuing advance of the desert margins into semi-arid regions. [6]

Main reason identified instantly!

Reason two.

The massive increase in population in the Sahara region of East Africa has led to a greater dependence on the land to provide both food and fuel. But this has proved too much for the already fragile ecosystem to sustain and much of the Sahara has been or is turning into desert. First is the need to graze livestock to provide food – the Tuareg tribe of Chad and Burkina Faso is one of the major perpetrators in the over-grazing of

Sample question and model answer (continued)

Reason three.

Shows a good grasp of the human interaction. This sort of case study knowledge is vital in all these types of question.

land. Their insistence on keeping larger herds than necessary has led to soil degradation as much of the vegetation is eaten away thus removing one of the key soil components. Second is the demand for fuel – the primary fuel source for cooking and heat in Africa is wood. It is not uncommon for young children of the Tuareg to be sent out to find pinewood as a day's task. This removal of trees and other flammable material combined with the increased over-grazing has seen the removal of much of the vegetation leaving the soil bare and hostage to the elements.

Comment

On the whole this was a super student response. There is clear and demonstrable knowledge. It more than fulfils the requirements of the questions. This individual went on to gain an 'A' in the subject.

AEB 1989

Practice examination questions

1 (a) Describe, using drawings, the principal features of the desert piedmont landscape. [20]

(b) Discuss the extent to which present day processes are responsible for the formation of desert piedmont landscapes. [25]

2 (a) Explain the reasons for the occurrence of tropical deserts such as the Sahara as a climatic region today. [4]

(b) Outline how movements in the position of the world's weather system could have led to greater precipitation in desert areas, such as the Sahara, in previous times. [6]

Assessment and Qualifications Alliance Specimen paper

3 With reference to the hot desert climate analyse the causes, and discuss the consequences, of its rainfall regime. [25]

Assessment and Qualifications Alliance Specimen paper

Chapter 4

Migration: current issues

The following topics are covered in this chapter:

- *The causes of migration*
- *The business of migration*
- *Limiting migration*

4.1 The causes of migration

After studying this section you should be able to understand that:

- *there are three main reasons why people leave their homes:*
  political – repression or civil war forces them out
  economic – they are forced to seek a livelihood elsewhere
  disasters – natural or man-made cause them to leave!

LEARNING SUMMARY

Around the world 1 : 120 people have been forced to flee their homes through war.

25 million migrants are the concern of United Nations High Commission for Refugees (UNHCR).

Eight out of 10 refugees are children and women.

Around the world there are in excess of 60 000 000 people on the move. Whatever the original impetus, migration is likely to become the big issue of the 21st century.

Definitions

Migrants: people who move from their home to another place. The migration may be internal or international. Migrants can return home if they wish to. There is no danger to their lives.
Refugees: Cannot return home or are in fear of their lives if they do so.
Asylum seekers: People seeking a place of safety after being persecuted in their own country.
Internally displaced persons: These migrants remain in their own country, but have been forced from their homes, usually by war.
Returnees: Refugees allowed to return to their own countries.

Persecution and conflict

AQA A	Some U5
AQA B	U4
EDEXCEL A	Some U5
EDEXCEL B	U4
NICCEA	Some U5

Population was covered in the AS book too. You should look at it!

Key points from AS

- **The dynamics of population**
  Revise AS pages 110–124

The endless conflict that is Afghanistan

During the Soviet occupation of Afghanistan in the 1980s and the civil war that followed, more than six million Afghans fled. Two million refugees have returned home since the Soviets left in 1989, another 2.2 million remain in Pakistan and Iran – including 340 000 who have fled since the Taliban seized power in 1996.

There are 311 000 Afghan refugees in UN camps and another 300 000 in local communities in Pakistan. Many are reluctant to return either because of the drought or fears of persecution at the hands of the Taliban.

A programme to voluntarily repatriate Afghan refugees in Iran is on-going and voluntary. More than 60 000 Afghans had been repatriated since the start of the UNHCR-sponsored programme.

Most of those Afghans were reluctant to return because they have health care and school benefits in Iran. In Afghanistan, most will not be able to attend school.

The UNHCR puts the total number of Afghans in Iran at 1.4 million. The Iranian authorities say there are 700 000 Afghans illegally residing in the country.

CASE STUDY

> Migrational flows currently have an enormous effect on countries; the country of origin and the host. It's a topical subject and is popular at A2.

There are currently some 60 000 000 refugees on the move, fearful for their lives. Generally they come mostly from countries in conflict.

Conflict such as that in Afghanistan tends to antagonise communities within affected countries and those countries that surround them, instead of drawing nations together, it forces them further apart. The fall-out from the refugee problem is an international problem that has very serious implications for both the developed and developing nations. What is new about the most recent refugee movements is the sheer size/scale of the movements. The map on pages 70–71 highlights fifteen such movements that have occurred, the graph below shows how refugee numbers have fluctuated over the years in the various regions of the world (figure 4.1).

Figure 4.1 *Number of refugees received in each region 1990–1998*

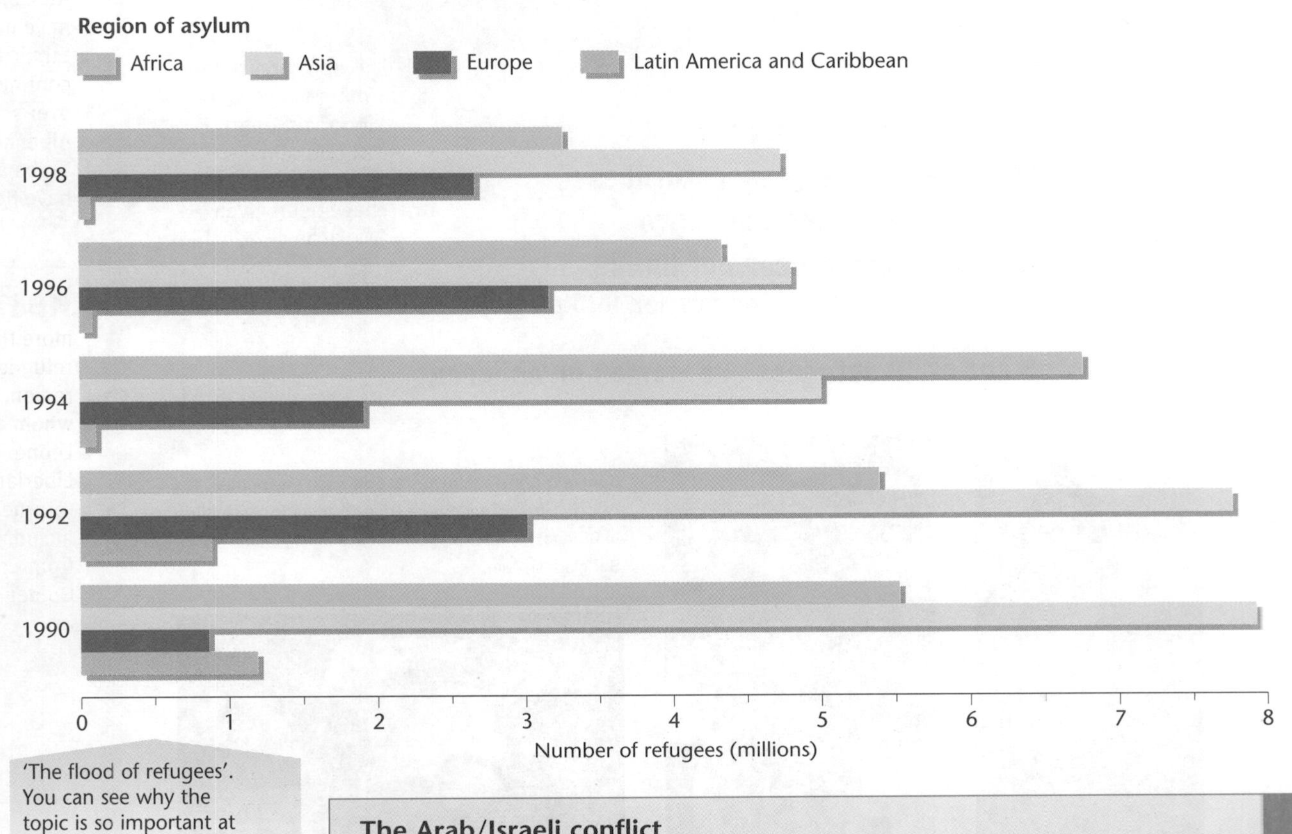

> 'The flood of refugees'. You can see why the topic is so important at A2 Level.

The Arab/Israeli conflict

CASE STUDY

Since 1948, some 3 200 000 Palestinians have been at the centre of the Arab-Israeli Conflict. Various treaties – the Madrid Peace Process, the Oslo Process, the Jordanian-Israeli Peace Treaty, for example – have all failed to find an amicable solution. For the Palestinians the 'right of return' is a central tenet, but the Israelis counter this by asserting they bear 'little moral responsibility for the plight of the Palestinians', and that *de facto* they accepted Jewish refugees from the Arab world.

Developments within the area over the last fifty years will undoubtedly impact on the way any return is handled. Many refugees may well leave one camp to be ushered into another!

The Israelis are keen to continue a form of demographic security.

The second thread of the Palestinian 'resolution' is the issue of compensation for their enforced exile. This amount totals some $40 bn to $92 bn.

Israeli opinion increasingly favours refugee resettlement, but in the refugees' current country of refuge or in 'third' countries.

Figure 4.2

focus on refugees

- **1 person in every 120 in the world today has been forced to flee his or her home as a result of persecution, violence of war**
- **22 million people across the world – refugees, returnees and displaced persons – are currently the concern of UNHCR** (United Nations High Commission for Refugees)
- **8 out of 10 refugees are women or children**

Guatemala 36,000 repatriated from Mexico since 1990, although 28,000 remain. Several hundred have been given Mexican citizenship.

Kosovo In 199 almost 1 million Kosovars (mostly ethnic Albanians) fled, although most have now returned. Ethnic tensions continue and now over 200,000 non-Albanians (mostly Serbs and Roma) have fled.

West Africa more than 600,000 refugees in the region, a third of whom are from Sier Leone. Over 250,00 Liberians have now returned home, including 180,000 from Cote d'Ivoire a Guinea.

© UNHCR

© UNHCR

from left: Afghan refugees at Sefid Sang reception centre; Liberian and Sierra Leone (West Africa) children in Guinea; Myanamar (Burmese) refugee woman in Bangladesh camp.

© Liba Taylor/Panos Pictures

Refugees are people who flee their country because of 'a well-founded fear of persecution for reasons of race, religion, nationality, political opinion or membership of a particular social group'. (UNHCR definition 1951)

Refugees either cannot return home, or are afraid to do so. Under the rules of the UN Convention a refugee has the right to stay in the new country for as long as may be needed.

Asylum seekers are people who have left their own countries claiming persecution and are seeking a place of safety. They may be granted refugee status in their host country and be able to stay.

Internally Displaced Persons (IDPs) like refugees, have been forced to leave their homes because of persecution, war or other threats, but unlike refugees, remain in their own country. Increasingly, they are the victims of civil war.

urope million have ught asylum since e early 1980s. ore are now oving to central d eastern areas.

Former Yugoslavia 3.5 million assisted by UNHCR during the 4 year conflict (2.7 million in Bosnia-Herzegovina)

Sri Lanka From 1992 - 95, 54,000 Sri Lankan refugees returned home from India (leaving another 64,000 there). 800,000 have been internally displaced as a result of fighting in the north of Sri Lanka.

Myanmar (Burma) 250,000 fled Myanmar in 1991/92 - 21,000 still remain in Bangladesh

The Caucasus 2 million displaced by a series of conflicts, unable to return because of continuing instability.

Horn of Africa the current home to 1 million refugees. 30,000 have returned to Somalia from Ethiopia, and others from the Sudan have been repatriated.

Palestine 3.2 million registered with the UNRWA (responsible for Palestinan refugees). Complex issue unlikely to be resolved quickly.

ngola Tens of housands fled in 1998 s fighting flared gain. About 10,000 ave been repatriated nd settled in areas ot affected by conflict.

Afghanistan 4 million Afghan refugees have returned home since 1990 but 2.7 million remain displaced (mainly in Pakistan and Iran). Fighting still taking place in some parts of the country.

Vietnam Since 1975 over 800,000 Vietnamese have sought asylum in S.E. Asia. 755,000 have been resettled and 110,000 returned to Vietnam via the UNHCR's CPA (Comprehensive Plan of Action). 2000 of these 'boat people' remain in camps (mainly in Hong Kong).

East Timor 250,000 (about a third of the population) fled the country in September 1999 after a referendum in which they voted for independence from Indonesia led to militia violence against the East Timorese.

entral Africa 1.3 illion Rwandans and 100,000 urandi returned home after the 996-97 war (in what was Zaire). till many refugees in the region .g. 250,000 Burundi and 45,000 ongolese are in Tanzania.

Courtesy of Worldaware www.globaleye.org.uk

Migrants are people who move from their home to another place. This may be **internal migration** – movement within a country – or **international migration** where migrants leave their country to live in another country, often seeking more money and a better life for their children. Unlike refugees, migrants are free to return home if they should wish to because, although they may be very poor, their lives are not in danger. These types of migrants are often called **economic migrants.**

Returnees are refugees who, when conditions allow, return to their own countries. They may return by themselves or with the assistance of the UNHCR or other agencies. This is called **voluntary repatriation.**

The Arab/Israeli conflict (outlined on page 69) is a complex issue, like all refugee situations. Further it is unusual in that normally the poorest countries of the world carry the burden of refugee situations. For instance, during Mozambique's 16 years of civil war, violence, torture and summary executions caused some 17 000 000 to flee into Tanzania, Malawi, Zambia, Zimbabwe, South Africa and Swaziland.

> **KEY POINT**
>
> **UNHCR (United Nations High Commission for Refugees)**
> - Set up in 1951 to tackle the problem of 1.2 million Ethiopian refugees left homeless after WWII.
> - Based in Geneva.
> - Works closely with Governments and NGOs to implement humanitarian aid.
>
> **UNHCR aims to:**
> - Protect and assist refugees, (i.e. ensure basic human rights).
> - Seek lasting solutions (i.e. return refugees or arrange asylum or resettlement).
> - Pay particular attention to the needs of children and the rights of women and girls (i.e. these are 80% of most refugee populations).
> - UNHCR's involvement may be long term, as in Vietnam, or brief (through QIPs, Quick Impact Projects) as in Mozambique.

The refugee challenge stemming from the second half of the 20th century seems likely to continue long into the 21st century. The hope is that conflict-driven migrants/refugees will decrease. It will require a change in the UN charter, however, for the UN to intervene within countries, to control conflicts, before people are forced to move out.

Economic pressures

AQA A	Some U5
AQA B	U4
EDEXCEL A	Some U5
EDEXCEL B	U4
NICCEA	Some U5

Increasing globalisation (the relative ease with which people can travel in the modern world) together with the wide and growing gap between rich and poor, is the catalyst, in many instances, for migrations based on the need for economic security. Within the world there are three foci, or interfaces, where most economic migration occurs, see below:

Figure 4.3

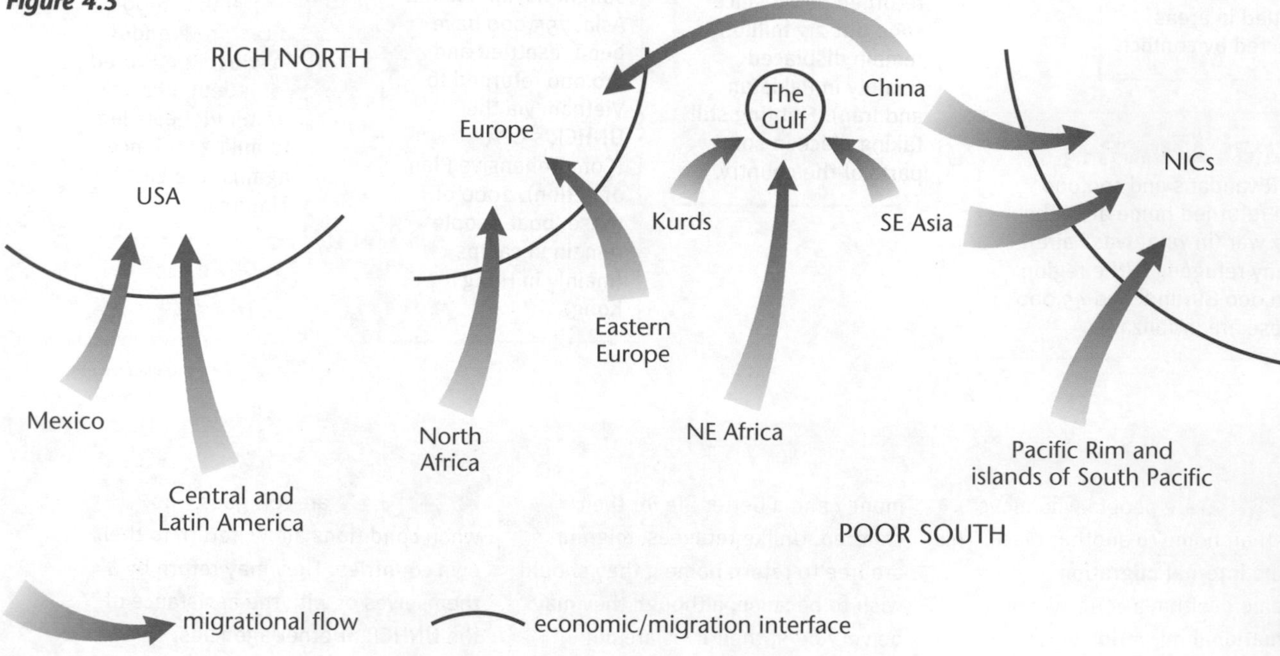

Interface 1: USA and Mexico

Open border 2001?

Mexico is keen to copy the model established in the European Union of open borders, to help 'create economic conditions which will help stem the flow of legal and illegal migration across borders'. The USA however, is not keen to adopt any model, be it the EU's or Vincente Fox's (the President Elect of Mexico), fearful that the US's wages and price of exports will be driven down, if there is a mass exodus through an open border! The USA, instead, is keen to develop Mexico's membership of the NAFTA (North American Free Trade Area) in order that both countries might benefit from the relationship.

Some facts on Mexican migrations

- About 7 million Mexicans live in the USA, equivalent to 8% of the Mexican population. Roughly a third are illegal immigrants.
- Mexicans easily top the lists for both legal and illegal immigration into the United States. In 1998, 131 575 Mexicans legally migrated to the USA, 20% of the total admitted to the USA as a whole.
- 2.7 million Mexican illegal immigrants, more than half the illegal immigrant total, are estimated to be living in the USA. The number is believed to be growing by about 150 000 per year, according to the USA immigration and naturalisation service.
- Two out of five illegal immigrants in the USA live in California. Texas, New York and Florida are the other states with high numbers.
- More than 1.53 million migrants, overwhelmingly Mexican, were detained by US border-control officers last year on the US–Mexico border. The vast majority, 1.49 million, were returned across the border.
- Last year 359 people died attempting to cross the border from Mexico into the US. Most deaths are due to drowning, or exposure and dehydration. Only about half the bodies are ever identified.

Source: The Guardian, August 2000

CASE STUDY

All these studies are well documented.
Read papers and browse the internet.

Interface 2: Europe and the Far East

Snakeheads are known in China as Sh'etan'. They are the leaders of the human trafficking trade. Snakes are the clandestine emigrants.

'Snakeheads and Snakes'

The growth of 'widow villages' in South East China suggests a sinister and high-staked gamble being played by many disgruntled, impoverished, rural Chinese. They are willing to pay up to £14 000 to buy their way out of China!

About 900 million people are linked to agriculture, but in fact only 250 million are needed! Hence the desire to 'escape' the country.

What many do not realise is the extent of the human trade and misery that is migration. Europe may be forced to open up its boundaries if the huge trade in people continues. At present some 400000 a year enter the EU from Eastern Europe and further afield, i.e. China. 21 000 were caught trying to enter the UK last year alone, this is 20 939 up on 1996!

On 23 June 2000 the terrible price that migrants pay for their 'economic freedom' was highlighted at Dover docks with the grim discovery of 58 Chinese dead in the back of a tomato lorry.

Professional migrants contributed $25 billion to the US economy between 1961 and 1972!

As a footnote to this 'interface', Britain announced in July/September 2000 that migrants with skills may be welcomed. Possibly the Government is at last viewing migration in context. Migration might relieve the economic pressures on overseas countries while, at the same time, addressing 'employment need' in this country.

A two-tier migratory system for Europe

The number of people of working age in the European Union is falling, many more immigrants will be needed to keep the economy running and prevent the collapse of the pension systems. Within 25 years, the EU will need up to 159 million immigrants to maintain the current ratio of workers to retired people. The twin problems in Europe are overall population decline and the unprecedented rate of ageing of the population. Population losses of 5 million are expected by 2025 and 40 million by 2050.

Britain has a positive birth rate and is closer to North America in its demographic structure, but even so it will face an imbalance early this century which will have to be offset by immigration or an overhaul of the retirement system.

KEY POINT

Interface 3: China and Hong Kong

There are an estimated 10 000 mainland migrants in Hong Kong seeking permanent residence. Originally the city-state's 'Basic Law' cast a wide net in granting Hong Kong residence rights to persons born in China, even if a person's parents were not Hong Kong residents when they were born. The law was reinterpreted in March 1999 to restrict Hong Kong residence rights to mainland Chinese with at least one Hong Kong parent at the time of birth.

Some of the migrants who would have been eligible under the first but not the second ruling are in Hong Kong and resisting a return to China. A militant faction last year set Hong Kong's immigration headquarters on fire to avoid being sent home, injuring 47 people. The arson attack has heightened anti-migrant feelings, and might further widen the gap between mainlanders and the residents of Hong Kong.

CASE STUDY

Natural and man-made causes of migration

AQA A	Some U5
AQA B	U4
EDEXCEL A	Some U5
EDEXCEL B	U4
NICCEA	Some U5

Natural disasters as a cause of migration

Compared to migrations brought about by political and economic reasons, displacements of population caused by natural disasters are small in number. Typically floods, earthquakes and volcanoes, drought, famine and climate change have moved people.

Being able to link the natural environment into your understanding of migration is vital.

Under the Volcano: Montserrat

The 5000 or so remaining residents of Montserrat, afflicted since July 1995 by the eruption of the Souffriere Hills volcano, are now settling down to the demanding task of rebuilding a society in the safe northern corner of their tiny eastern Caribbean island. Two-thirds of the island remains out of bounds still threatened by the volcano. Abandoned communities have become ash-ridden ghost-towns or, worse still, have been buried under millions of tonnes of volcanic rubble. The capital, Plymouth, has been virtually destroyed.

During the crisis, perceptions in Montserrat took on a colonial model of mistrust. At the worst moments, many Montserratians believed that the UK was deliberately seeking to 'depopulate' the island, either completely or sufficiently enough to reduce Her Majesty's Government's financial burden. The British always denied this, maintaining that given its priority to protect lives it had to make contingency plans for a worst-case scenario.

There remains then uncertainties of the still-active volcano, which continues to cast its shadow over the lives of Montserratians, both those who moved across the island and those who have moved away.

CASE STUDY

CASE STUDIES

El Niño and Peru

The effects of El Niño can be catastrophic. Floods, droughts, hurricanes and other weather-related phenomena cause immense loss and suffering. The Andean regions of South America have been especially hard hit due to the large scale interaction between the flow of water off the west coast of south America and atmospheric winds, oceanic temperatures and currents.

In 1998, in Peru alone, damages were estimated at US $3480 million. Gross Domestic Product faltered by 12.3%; 2634 kilometres of roads were destroyed and 50% of the cultivated areas in the departments of Tumbles and Piura (Northern Peru) were lost. On the high plains, drought affected 70% of the cultivated area, which resulted in migration and depletion of the population.

Somalia: flooding

In 1998 a combination of severe flooding and crop failure, the Gu, which accounts for 80 per cent of the country's annual cereal production, produced the poorest yields in five years and led to Somalis moving *en masse* to the coast. When people start to migrate in this way in search of food, we know a critical level has been reached.

In response to the emergency, 500 tonnes of maize and lentils were moved from the port of El'maan in southern Somalia for 100 000 people, at a cost of $1.2 million.

KEY POINT

Global warming and migration

There have always been extreme weather events, but the number and severity has increased in recent years. Severe weather events have claimed between 50 000 and 100 000 lives since 1997. In addition, up to 300 million people have been displaced and made homeless.

Climate disturbances will lead to mass migration as people move from affected areas. Disruption of agriculture and economies will lead to an increase in poverty. Disease will increase as viruses become more prone to jump species barriers.

'Climate disruption will take at least 100 years to play through – it takes that long to see permanent irreversible change as a result of climate change.' There has been an increase in population movement across borders in West Africa as a result of drought. In East Africa, Mozambique's economy has been set back years by massive flooding and consequential migration.

The rise in extreme weather conditions guarantees that these problems will get worse. A major inundation in Bangladesh, for example, could displace millions of people, and put the Bangladesh–India border under pressure. The result: increasing conflict. 'When you get scarcity of resources, population movement and poverty, you almost inevitably end up with conflict and destabilisation. It will make the world a more conflict-ridden place.'

Man-induced migrations

Failure of agrarian economies

Lack of resources can also drive people to migrate. Deforestation and soil damage have forced many Africans to alter their agrarian life style and go in search of waged labour to supplement the family income.

Africa has 12 million displaced people and half are supported by UNHCR.

The high prevalence of HIV in Africa has many causes – the principal one is probably migration! 40% of migrants contract AIDS.

Foreign workers send up to $250 billion home each year to their country of origin.

CASE STUDIES

Mozambiquans in search of prosperity

- Every year, South Africa expels about 100 000 Mozambiquans. South Africa is much richer than most of its neighbours. Wages do not, of course, compare with those in Europe or America, but South Africa is easier for its neighbours' citizens to get to. South Africa's land border is roughly 4 000 km long and poorly guarded. Crossing is not always safe: migrants are sometimes eaten by crocodiles as they swim the Limpopo river, or mauled by lions as they creep through the Kruger Park.
- South Africa has long played host to multitudes of migrants, and estimates for the number of undocumented migrants in South Africa range from 2 to 8 million in a local population, which is about 46 million strong.
- The newcomers are not always popular. Many locals fear that immigrants will crowd their towns, steal their jobs, and corrupt their culture. Hostility towards migrants sometimes flares into violence. South African hawkers have destroyed foreign-owned street stalls and beaten up their owners for selling goods too cheaply.
- South African mines, for example, employ about 120 000 foreigners, mainly from Mozambique and Lesotho. Foreigners are 'more skilled, more productive and less militant'.
- Remittances from South Africa make a big difference to neighbouring economies. In Lesotho, miners' remittances account for almost a tenth of GDP, although this wobbles with the gold price.
- In southern Mozambique, almost three-quarters of respondents to a survey said that a family member was working in South Africa. Among those who had been to South Africa themselves, 87% had saved enough to buy a house.
- Mozambique would be poorer, if its people could not work and trade in South Africa.

Haitian migration

Neither the threat of deportation nor the high risks of the Atlantic voyage to the USA has deterred thousands of Haitians from trying to escape deteriorating political and economic conditions.

The wealthier would-be-migrants head for Florida. They pay £2500 for the trip, but face automatic deportation. Cubans however, as a legacy of a cold-war-era law, are permitted to remain in the USA pending an immigration hearing if they manage to reach dry land.

No-one is sure how many Haitians disappear without trace each year attempting the perilous sea crossing. But the exodus is increasing.

Development causes migration

Infrastructural projects also cause massive displacement/migration/re-location. Globally 12 million a year are 'moved' by such new projects.

CASE STUDY

Dam resettlement programmes and the Three Gorges Dam

Human rights violations associated with the displacement of people for the construction of massive dams is a growing yet neglected problem. An estimated 30 to 60 million people world-wide have been forcibly moved from their homes to make way for major dam and reservoir projects. These 'reservoir refugees' are frequently poor and politically powerless; many are from indigenous groups or ethnic minorities. The experience of more than 50 years of large dam building shows that the displaced are generally worse off after resettlement, and more often than not they are left economically, culturally and emotionally devastated.

Man actually forces migration on himself!

CASE STUDY

Now China is planning the largest ever such relocation: the movement of between one and two million people to make possible the Three Gorges Dam, the world's largest hydroelectric dam. This displacement could also turn out to be one of the world's worst reservoir resettlement disasters. Unlike some of their counterparts around the world who are successfully mobilising to challenge massive dams and to defend their right to their land and livelihoods, Three Gorges resettlers await their fate mostly in silence, their concerns censored out of media reports and concealed even from the eyes of central government officials.

Owing to the inadequacy of financial and material resources allocated for resettlement, they fear it will be virtually inevitable that there will be major confrontations between people to be relocated and the authorities.

4.2 The business of migration

LEARNING SUMMARY

After studying this section you should be able to understand that:

- *people have replaced drugs as the most lucrative trade item for criminals*

Moving people for profit

AQA A	Some U5
AQA B	U4
EDEXCEL A	Some U5
EDEXCEL B	U4
NICCEA	Some U5

People want to be 'safe' and many want a better life. As a result there is a growing trade in people trafficking. Hundreds of thousands want to 'escape' and are prepared to pay huge sums of money to achieve this. They may have to take a clandestine route to safety because of the exclusionary measures that have been put in place in the host countries: draconian visa regimes and penalties imposed on airlines and lorry drivers harbouring immigrants. For many, illegal entry into other countries is the only option. The trafficker becomes the migrants lifeline!

Each migrant, or snake, had paid the equivalent of £14 000 to get to the UK.

400 000 entered the EU illegally between March 1999 and 2000. 21 000 entered the UK in the same period. The figure in 1996 was 61.

Could it be that criminals are reacting more quickly to changing patterns of labour demand, ahead of governments?!

Often the people who deal with 'the paperwork' are ruthless criminal gangs, witness the 58 Chinese who suffocated to death last year in the back of a refrigerated truck, and the dozen or so who hitch lifts in the wheel bays of jet airliners each year! Migrants keep strange, ruthless and brutal company in their desire for sanctuary.

With criminal gangs involved, failed immigrations are common, often leaving the migrant penniless and in debt. Few 'home' countries worry about this illegal trade however as 'illegals' contribute to GNP by sending money home. Those that fail to secure employment in the 'host' country frequently end up involved in crime, usually to pay back money for tickets and passage.

CASE STUDY

Italy and fortress Europe

Italy has a reputation throughout Europe for being the weak spot in the fortifications of 'fortress Europe', a title it would be only too happy to lose.

According to the Interior Ministry, immigrants from non-European Union countries legally present in Italy numbered 806 027, in 1998. The majority of these were from Morocco; 119 381, followed by 73 126 from the former Yugoslavia, and 70 897 from Albania.

The non-documented illegal immigrants number around 236 656. These immigrants permeate into all corners of Europe, helped on their way by the criminal fraternity.

Ceuta, then Europe!

Many Nigerians are prepared to cross the Sahara in search of a job. Their goal is Ceuta, a Spanish enclave tucked into Morocco's Mediterranean coastline – a hike of about 3000 miles. They hire the services of a trafficker, who takes their money, puts them in a lorry and dumps them at the border.

Spain has hung on to Ceuta and its twin enclave, Melilla. On Spain's entry into the European Union in 1986, two forward posts of the EU came into existence in Africa. The trickle of people began in the mid-1990s. In 1997, about 700 'illegals' came up through the Sahara and entered Ceuta from Morocco. By 1998, the figure was nearer 1000. Last year it was 7000. The year-on-year increase projected on these figures is worrying, to both the Spanish authorities and the EU.

Ceuta is another vulnerable flank of 'Fortress Europe' – a tempting point of transgression where migrant pressure has to be opposed. In 1993 the EU approved funding for a defensive wall round the enclave. It has two high parallel wire fences running for eight miles round the territory, a roll of razor wire, a line of electronic sensors and 33 closed-circuit cameras. The cost was estimated at $25 million. The game is on: a contest between the low technologies of willpower and mutiny, and the hi-tech security skills of the rich world.

By 2015 it is estimated between 15 and 20 million migrants will have made a bid for Western Europe via Spanish territory.

Dealing with clandestine migration is high on the EU's agenda. It's suggested 1.5 million migrants enter the wealthy European states every year and that 'every other migrant in the "first world" is there illegally'.

CASE STUDY

For the EU the problem appears to be insurmountable; frontiers are becoming harder to penetrate; but it is unlikely the trade in humans will stop. As long as there's money to be made a trafficker will be there to take it from unwitting 'migrants'.

4.3 Limiting migration

After studying this section you should be able to understand:

- *what can be done to prevent destabilising movements of refugees and immigrants*
- *that most people want to stay in their own countries if there is no persecution and a reasonable standard of living*

LEARNING SUMMARY

Restrictive measures

AQA A	Some U5
AQA B	U4
EDEXCEL A	Some U5
EDEXCEL B	U4
NICCEA	Some U5

Tighten asylum rules and fortify Europe?

This is hard to implement against a backdrop of open borders in Europe, but it is a policy that many favour.

In 1980 there were some 70 000 asylum applications made in Europe. In 2000 this had risen to about 800 000. Germany was the main target, but like the UK, France and Spain, Germany rejects nearly 92% of all applications. In fact several governments have come to power in Europe with tacit support from an electorate warned about rising immigrant numbers and potential home unemployment!

If it is accepted that large-scale cross-border legal/illegal migration is a fact of life and here to stay, it is likely that the only effective method of managing the situation is to reduce the motivation to migrate in the first place.

Resolving conflicts

Displaced migrants, in the first instance, probably want to live in a peaceable environment. War is one of the biggest causes of displacement and one of the hardest to resolve.

Governments are only too aware of the affect migration can have. They 'plan' to deal with the migrational flow.

Return to Kosovo

NATO, *'we will fulfil our promises to the Kosovar people that they can return to their homes and live in peace and security', and also pledges 'the deployment of an international military force to safeguard the swift return of all refugees and displaced persons'*

Patterns of prolonged displacement elsewhere in the Balkans and the rest of the world have taught some lessons about the standards necessary for people to return to their homes – and return for good.

Security
Refugees say that the most important condition for their return is safety. In any post-war situation there are many different security threats. Weapons, ammunition and explosives are rife, and a disorganised society creates space for all kinds of criminal acts. Milosevic's army has planted thousands of mines in strategically important areas, and mines have a strong psychological impact alongside the physical. Another security threat is the real or imagined fear of harassment, and the uncertainty of the future relationship between the Kosovar Serbs and the returnees.

Documentation
Closely linked to the security question is the issue of personal documentation and the documentation of property. Kosovars were deprived of their personal documents before being forced out of the country. To issue new documents is in itself a time-consuming and difficult job, involving the restoration of rights and assets that may be disputed.

Shelter
Many houses were destroyed; reconstruction is unlikely to be any swifter in Kosovo than elsewhere.

Infrastructure
Fighting, sabotage by Milosevic's army and NATO's bombing has destroyed much of Kosovo's infrastructure. This will be slow and costly to rebuild.

Income and social security
Even when security, documentation, shelter and infrastructure are in place, people cannot return to a place where there is no likelihood of economic survival.

Prospect of lasting peace and commitments
If there is little prospect of a lasting peace, few people will opt for repatriation.

Reconciliation
Possible internal conflicts among the returning population can also create problems.

Motivation and information
Correct, reliable, concrete and comprehensive information must be made available to refugees, IDPs (Internally Displaced Persons) and all those involved in relief and support efforts. The existence of accessible, independent media can make a decisive difference compared to a situation where IDP's refugees and returnees are left victims of propaganda, misinformation and prejudice.

Time
The longer people are away from home, the more they integrate with and get accustomed to their new environment.

CASE STUDY

CASE STUDY

Distance
The further away refugees have been settled from their homeland, the less chance there is for voluntary repatriation. For the many thousand Kosovar Albanians who have been and will be evacuated to places far away from the Balkans, talk of voluntary repatriation is likely to be more rhetoric than reality.

Overcoming trauma
Going back to the place where atrocities occurred is extremely difficult for some people.

Reducing migrationary pressure

Sustainable development that includes and encourages education, healthcare and sanitation, along with the provision of jobs, would greatly reduce migrationary pressure. Restoration of democracy is also important.

Increasing tolerance

If 'host' countries can be encouraged to accept that 'immigration strengthens countries', illicit trafficking would be lessened and refugees more easily rehomed.

Summary

The biggest migrationary problems are the result of complex historical, racial and political conflicts and issues. Law and order are increasingly difficult to accomplish in increasingly lawless societies and developmental aid doesn't assist the process.

Migrational crises will continue until:

- stable government is universal and
- economic migration is halted.

The South has to catch up with the North! It is unlikely we have yet seen migration at its height or at its worst!

Sample question and model answer

1

'International migrations have changed a great deal in the last 30 years, but these changes have had little effect upon the issues that arise in multi-cultural societies which have generally stayed the same.'

Discuss the extent to which you agree or disagree with the above statement. [25]

Good definition.

International migration is the movement of people across the border of a country, leaving one country for another. As with other types of migration this is of two types, voluntary and forced migration. Voluntary migration can further be split into permanent and temporary migration.

Identifies the changing levels of migration.

Permanent voluntary migration includes migration of people from a developed country to a developed country, a classic example of this is UK doctors moving to the USA for more money. This type of migration has probably not changed much within the last 30 years. It may even have increased with the increasing unity of the EU that makes it easier to move between European countries and to find work. Improved transport links, particularly with flying, will have made permanent migration a more viable option. The increase in over 65s in populations such as the UK could also increase international permanent migration as the retired and over 65s more to warmer climates such as Spain and other Mediterranean countries.

A range of levels and stages of development offered.

Good appropriate paragraphing.

The second type of voluntary permanent migration is from a developing country to a developed country. This type of migration was very common 30 years ago as countries tried to supplement their workforce with cheap labour. This happened with Mexicans migrating to California in the USA and also Algerians migrating to France. Recently such migrations have become less common as the developed countries have encountered increasing unemployment and cheap foreign labour is no longer required, indeed the migrants face racial tension. Because of the lack of employment, the MEDCs tightened up immigration rules and make obtaining work permits harder. This type of migration is also decreasing because if the economy of the MEDC declines, it is the migrants who tend to lose their jobs first and are hit hardest. This has happened to the Turkish people who were migrating to West Germany. Now many of these people are moving to Saudi Arabia and Libya to take advantage of jobs created by oil.

Accuracy?

Deals with issues.

To be an issue, two positions have to be identified. This happens here.

Finally there is international migration from LEDC to LEDC. This tends to be small and relatively unchanging. This type of migration often happens when oil is discovered. In Venezuela oil was discovered and people from neighbouring South American countries such as Colombia migrated into Venezuela to take advantage of the oil jobs.

How changes affect issues.

Voluntary temporary international migration has the same major ways of migration. First is MEDC to MEDC and this is again probably increasing slightly as people migrate for short periods for holidays or jobs. This is increasing because international transport is becoming quicker, easier and cheaper particularly around the EU. The increasing internationalisation of major companies is increasing this type of migration.

Well aware of what drives current migration.

Migration from an LEDC to an MEDC on a temporary basis is decreasing for the same reasons as before with stricter immigration laws and difficulties in obtaining a work permit. It is also becoming more expensive for people of an LEDC country to move to an MEDC and is too expensive for only temporary migration.

Sample question and model answer (continued)

Avoid too much of your own opinions. All ideas have to be substantiated.

Migration on a temporary basis from LEDC to LEDC is difficult to quantify in terms of change over the last 30 years as figures for such migration are limited. I would have to say that this migration has probably increased where the economic and development gap between LEDCs has increased. This could happen where a country discovers an important resource which neighbouring countries lack, however people migrating to the country could migrate permanently unless transport between the countries is good and over a short distance. Between LEDCs transport is not good so most migration between LEDCs will be permanent.

Issues drive this area of geography.

All aspects that are relevant and pertinent have been dealt with.

The final type of international migration is forced migration. It is hard to say how this type of migration has changed over 30 years as it depends on fluctuating factors. Some years this forced migration will be largely thanks to a war flaring up: e.g. in Africa, war in Sierra Leone resulted in many refugees. Other years may be more stable and such migration will be limited. Other types of forced migration such as when caused by famine could be increasing as increasing pressure is put on food resources by growing populations.

Where?

Was a decent conclusion not necessary?

International migration is often a source of problems as people of different cultures, races and with different religions mix. In many countries, migrant foreigners are treated in an unfair way and experience racism. The problem is that people of the host country view the migrants as job stealers and see them as taking advantage of government-provided social services. The migrants only want to improve their standard of living. This issue may have decreased with decreasing levels of migration particularly to MEDCs. Unfortunately the race issue continues to affect second generation migrants born in the host country. The migrants do little to warrant such racism and often keep to their own communities but still experience racism. This can follow migrants as they attempt to better themselves with better jobs and housing. Second-generation migrants may be denied equal access to jobs. Meanwhile the indigenous population sees migrants taking their jobs and by accepting lower wages making their own wages lower. Because migrants work for lower wages indigenous people may lose their jobs first and those unemployed resent the working migrants further. In theory, as migration declines so should this problem but the second generation still experience the problem so it is not going away.

Comment

Overall: a good knowledge of the migration process at a variety of levels.

This essay addressed, in a very structured and organised way, all the issues that were pertinent. What it didn't do was address them with up-to-date case studies. Mentioning countries is fine, but to demonstrate how the detail of actual places fits into what is being discussed is a must.

Practice examination questions

Some tips below are offered to help you answer the question more accurately.

The diagram shows a population structure greatly influenced by migration. Some towns and regions within countries of the **more** economically developed world have this type of population structure.

Ensure you always use resources that are offered!

Look especially at anomolous age groups!

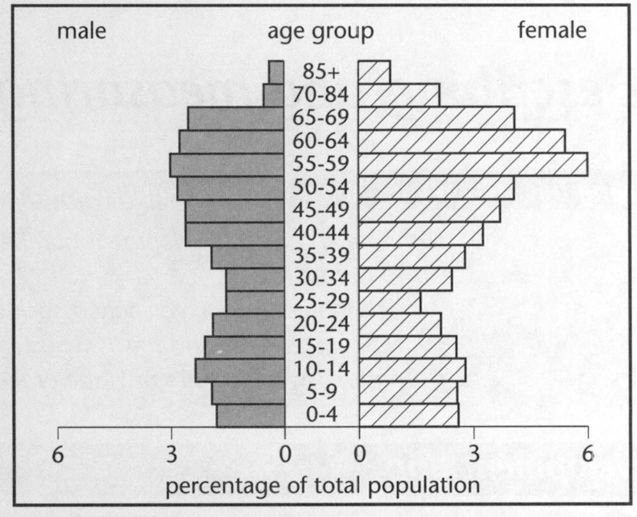

Source: adapted from B. Knapp, Systematic Geography *(Harper Collins) 1992*

Typically as an A2 question 'explain' occurs more frequently than 'describe'.

(a) Explain what kind of town or region this could be if the dominant type of migration was:

(i) in-migration

(ii) out-migration. [6]

Attempt to offer as many ideas, that are relevant, as possible.

(b) If the population structure shown in the diagram resulted from in-migration, outline the possible economic benefits for this town or region. [6]

If the question asks for specifics 'supply' them! You must include a real example to reach the highest marks!!

(c) For a named region or country into which people have migrated, explain the unfavourable consequences of migration. [8]

[Total: 20]

Assessment and Qualifications Alliance

Chapter 5

Development issues

The following topics are covered in this chapter:

- Describing and measuring development
- Spatial differences in development
- Causes of disparity
- Aid and trade as forces for change
- Inequalities in MEDCs

5.1 Describing and measuring development

After studying this section you should be able to understand that:

LEARNING SUMMARY

- there is a wide development gap between the Northern and Southern Hemispheres
- development is increasingly associated with human welfare and not just economic gains, and this is reflected in the criteria chosen to describe a nation's wealth/state or stage of development

The development gap

AQA B	Some U4
EDEXCEL A	U5
EDEXCEL B	U5
OCR A	Some U5
WJEC	U4

Development is difficult to define. Is it about human welfare, environmental sustainability or economic growth?

As an informed A2 candidate you might argue for any or all of these.

The development gap is commonly examined at A2.

The richest 10% in the world control 60% of the world's wealth.

The last three chapters of the AS book support many of the issues in this chapter.

Remember to access all the knowledge you can.

Key points from AS

- **Settlement issues** *Revise AS pages 90–106*
- **The dynamics of population** *Revise AS pages 110–124*
- **Worldwide industrial changes** *Revise AS pages 126–138*

80% of the globe's population struggles to develop the resources that lead to a dignified and productive life. This struggle is often called 'development'! This struggle is also viewed as a process of change operating over time. You will be aware of the fact that the countries of the world have not developed equally. The focus of much of the development work carried out in the latter part of the last century and well into this new century will be to ensure that regions develop equitably, with adequate food supplies, medical services and educational opportunities, under regimes that value social justice and political and economic freedom, and quality of life. Sustainable development, too, will be high on the agenda.

Development is then a complicated and complex process driven by many variables. On the one hand the problem is economic, on the other it is political, environmental and social. It is a process operating in the context of growing disparity between nations and within poor nations, the so-called **development gap**.

Figure 5.1 *The development gap*

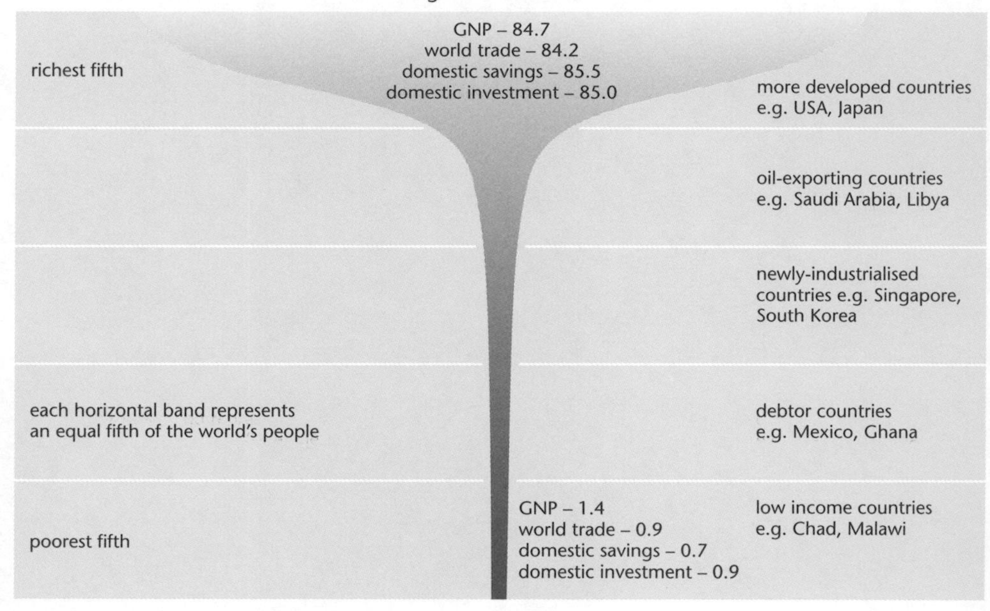

Source: Human Development Report; UN

Describing development

AQA B	Some U4
EDEXCEL A	U5
EDEXCEL B	U5
OCR A	Some U5
WJEC	U4

The term development is a relative term since a country is only developed or developing by comparison to others.

Nations with high living standards are said to be developed – they have gone through the trials of development. Those that currently negotiate the barriers to development are said to be developing. To Western observers industrialisation holds the key to development, hence the tag Less Economically Developed Country (LEDC). The industrialised West being the More Economically Developed Countries (MEDCs)! In the very near past the poor nations have also been 'labelled' the third world, the under-developed world and so on!

Some 23 years ago Willy Brandt was tasked by the UN to both clarify and to suggest solutions to global inequality. In his report 'North South a Programme for Survival' an urgent plea was made for change: 'For peace, justice and jobs'. His report also established the North v South divide of the world, see below:

Trends since the Brandt Report

Figure 5.2 The Brandt line dividing the North from the South

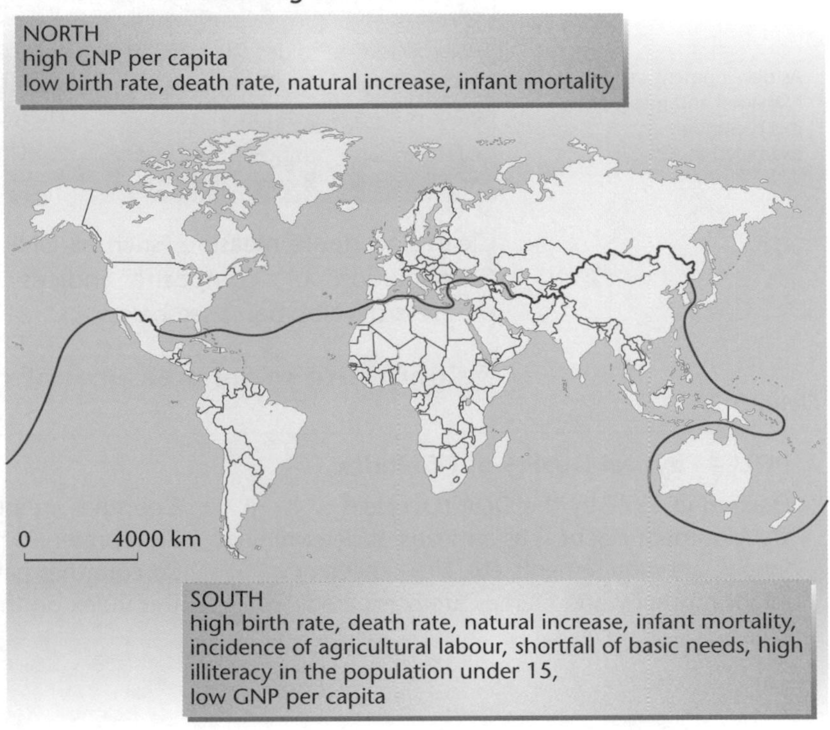

Positive

- Many countries have increased production and standards of living have improved.
- The global economy has strengthened.
- The cold war has ended.
- Military expenditure and arms sales in/to the south have declined.
- Democracy is supported around the world. Colonialism has all but vanished. Apartheid has gone.

Negative

- Absolute poverty has increased: 1.4 billion live on $1/day.
- Children continue to die through hunger.
- Population growth is not under control.
- Environmental threats are becoming more alarming.

The poorest nations are largely tropical in nature. From the Sahara to Malawi and Ethiopia to E. Asia.

Much has changed since the Brandt Report, but just five years ago the Commission on Global Governance warned that the imbalance between rich and poor, and the lack of stability in managing the system, risked 'destabilising' shocks that would rock the world.

Measuring development

AQA B	Some U4
EDEXCEL A	U5
EDEXCEL B	U5
OCR A	Some U5
WJEC	U4

In the past to measure development was easy, economic indicators being the sole indices used. **GNP (total value of goods and services produced by the country divided by population)** was the principal indicator. Nowadays a number of broader alternative measures can be used, and rightly so, including measures to do with social well being, political and material well being. The table below charts the limitations of using GNP as the sole measure of development and offers a range of alternatives.

Within nations, individuals have access to wealth and power and privilege.

Costa Rica is an LEDC based on its GNP. By any other measure it's not!

Limitations of GNP as a measure of development:	***Additional indicators of development:***
Accurate data is hard to find/acquire	% workforce in agriculture
Currency rates vary on a daily basis and it is difficult to put an accurate value on goods	% population increase Life expectancy People: doctor ratios
Subsistence industries and farming production do not enter the market economy and are therefore not measured	% with access to safe water % unemployed Telephone boxes per 1000 of population
Per capita GNP hides the internal distribution of wealth in a country	Calories per capita per day HDI (Human Development Index)
GNP is always given in $US	Literacy rates
GNP reveals nothing about quality of life and well being (ditto GDP)	% school enrolment as a % of relevant age group The status of women

As development can be managed and measured it is commonly examined at A2.

Clearly no single measure, such as GNP, can adequately indicate development. Combination or composite indices really are far more illustrative of progress/development, see below:

Composite social measures of development

Figure 5.3

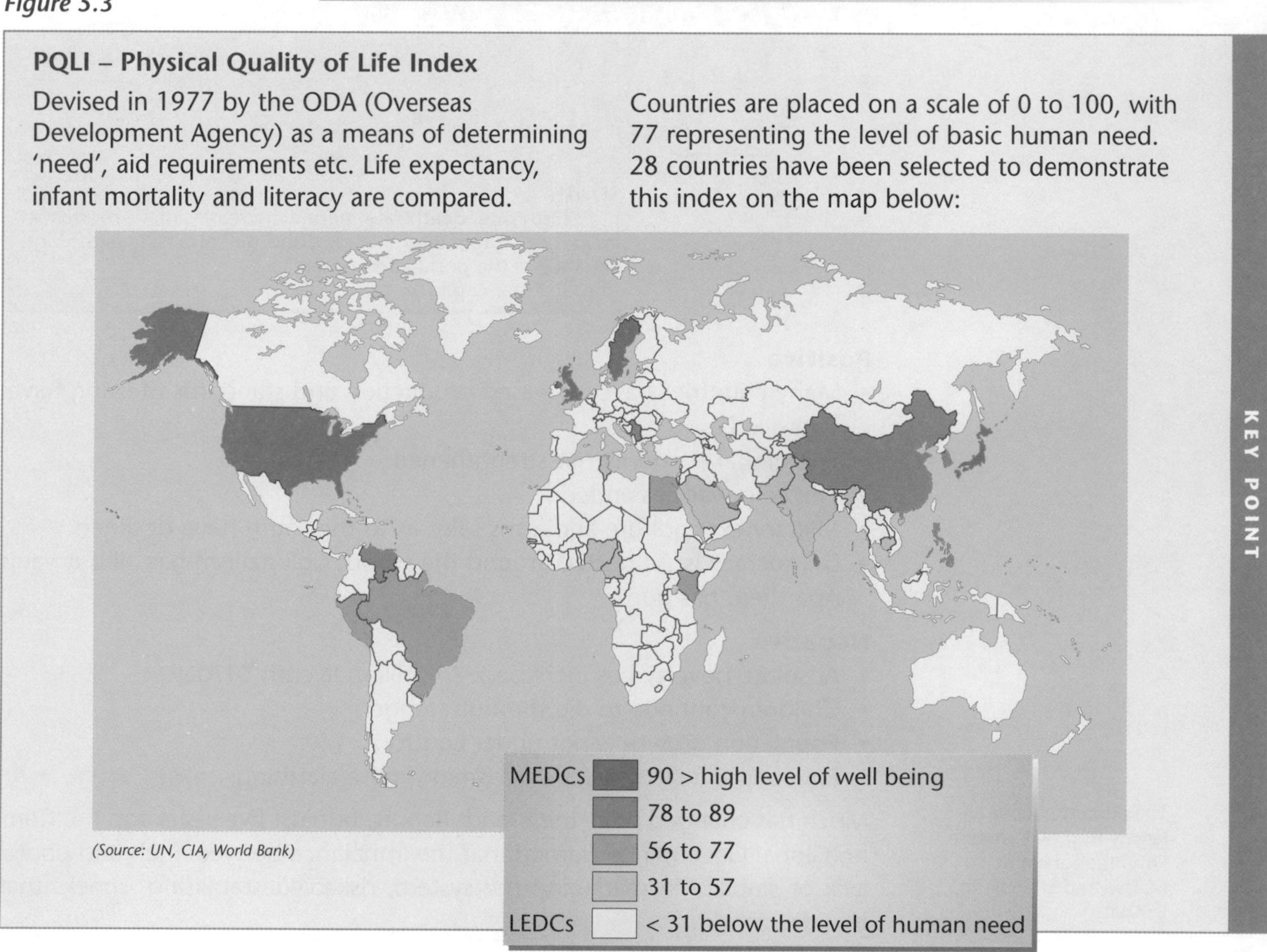

HDI – Human Development Index

Devised by the UN in 1990, it uses three development indices to assess development: real per capita income; a measure of adult literacy; and life expectancy at birth. Countries are placed on a scale from 0 to 1: the nearer they are to 1 the more developed a country is.

Problems with social measures of development

- They do not reflect inequalities in income distribution
- There is no agreement over which social indicator to use
- Rights of women and freedom of speech cannot be included
- Development is about inter-dependence between countries, how can this be measured?

KEY POINT

5.2 Spatial differences in development

After studying this section you should be able to understand:

- *that models of regional development help explain how regional inequalities occur*
- *many problems facing LEDs have their roots in their colonial past*

LEARNING SUMMARY

Theories of development

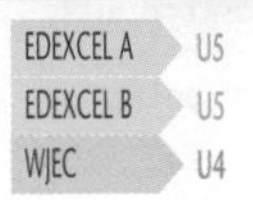

Explanations of how the MEDCs have become rich are many and varied.

Rostow's development stages model

The American economist W. W. Rostow argued that the countries of the 'North' passed through a series of developmental stages and that those in the 'South' are now passing through these stages.

Figure 5.4 *Rostow's development stages model*

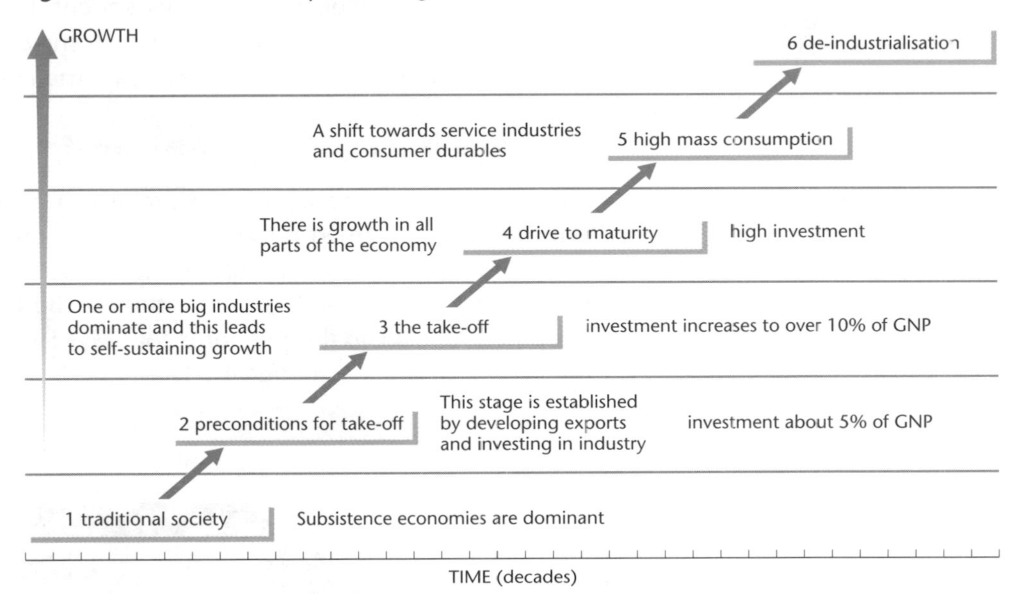

Limitations

- Eurocentric in origin – i.e. it is assumed the MEDCs, particularly Europe, have all the answers.
- How accurately can the economies of the twentieth century 'North' be compared effectively with say Guyana today?
- Cash offers and injections have failed to bring about take-off in many LEDCs.
- Late developers do not have the resources that were available to the developed countries when they industrialised.

Frank's dependency theory

'The development of underdevelopment' is a well-worn statement in development circles, one recognised by André Frank, who believed the MEDCs (of the North) have insidiously bled the LEDCs (of the South); exploiting the people and the resources. His argument originates in the sixteenth century (the so-called colonial period) and explains why the developing world became poor and has stayed poor.

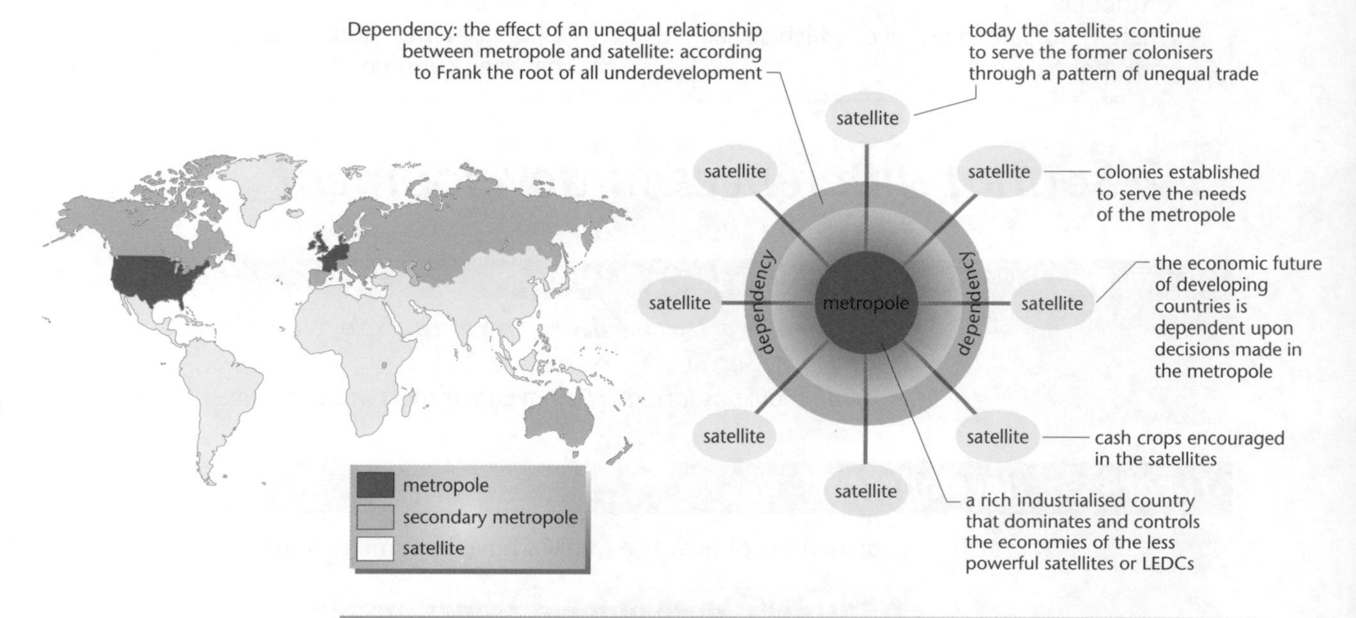

Figure 5.5 Frank's dependency theory 'The colonial impact'

Blaut and Neo-colonialism

James Blaut used the term neo-colonialism to describe the relationship between developed and developing countries at the present time in his book 'The Theory of Development'. To his mind, though LEDCs are politically independent of the MEDCs they are still economically dependent. Banks, investment, aid, exports and trade are all controlled by the MEDCs. In some countries the MEDCs still play a big part, e.g Sierra Leone – the UK, Rwanda – the French and Dutch and so on.

Benefits are brought by colonialism, i.e. roads, rail, health services and schools.

Economic dualism – Myrdal and Friedmann

Models are representations of reality. Examiners love them!

The colonial era encouraged the growth of 'modern', expanding and capitalist economies that formed pockets or cores of commercial activity within the much larger traditional economy. As the core areas enlarged, the economically weak traditional areas shrank in importance. Both Myrdal and then Friedmann attempted to explain this phenomenon. Myrdal attempted to explain and decipher the causes of regional inequality. Thus:

Figure 5.6

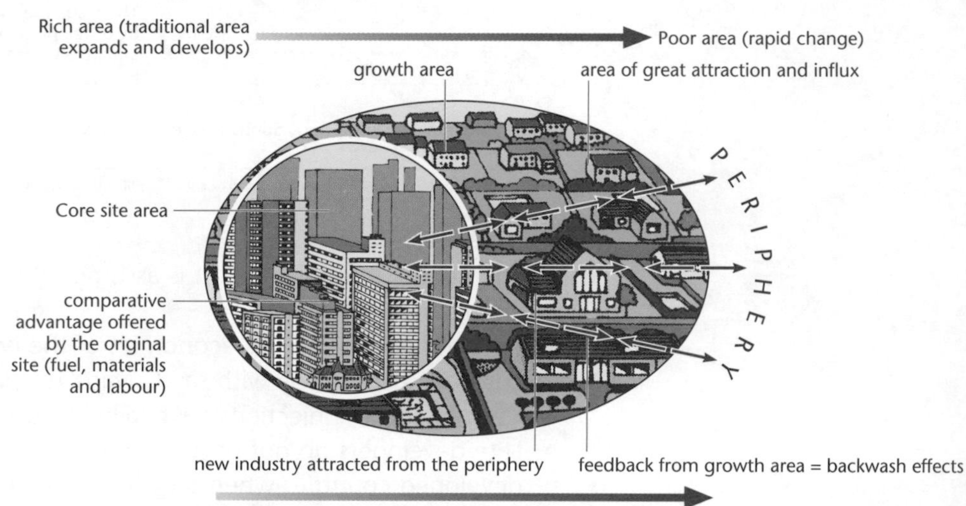

Friedmann explained

KEY POINT

Figure 5.7 Friedmann's development model

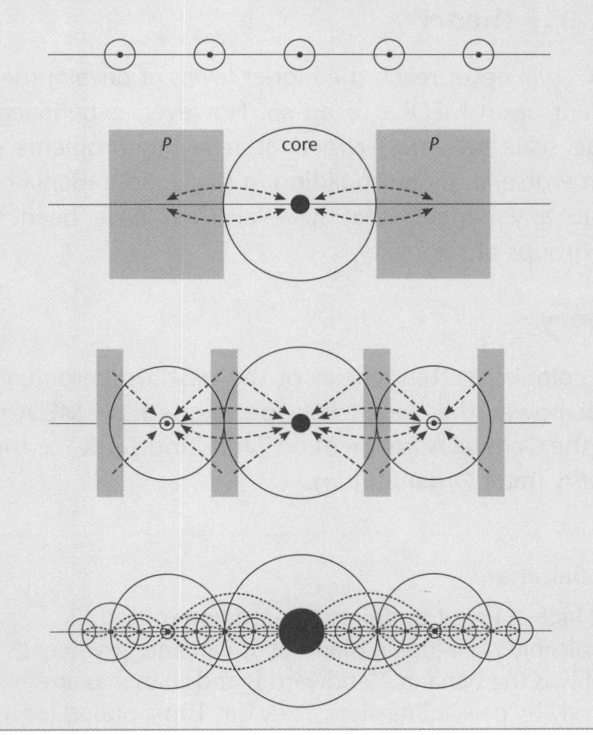

Stage One A series of independent local centres with no hierarchy. Each town lies at the centre of a small region – typical pre-industrial structure.

Stage Two A single strong core emerges, together with a periphery from which potential entrepreneurs and capital migrate. The national economy is reduced to a single metropolitan region – typical of incipient industrialisation.

Stage Three There are now strong peripheral sub-cores as well as a single national core. During industrialisation, secondary cores form and in so doing subdivide the periphery.

Stage Four Here there is a functional interdependent system of cities, efficient in location to give maximum growth potential.

CASE STUDY

Figure 5.8 Friedmann – as exemplified by Venezuela

With Caracus as the Core: an urban industrial area with high levels of technology and capital investment. (Various multipliers and agglomeration allow the core to grow.)

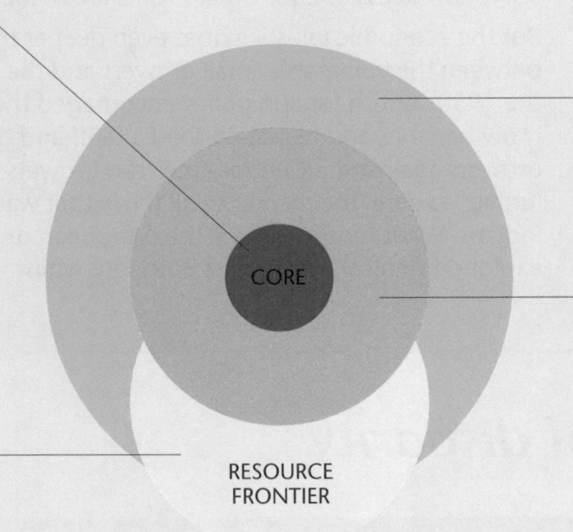

The Venezuelan western mountains become the **downward transition region**: a stagnant or declining industrial region or backward rural region.

The Valencia Basin in Venezuela becomes the **upward transition region**: an area of economic growth and spread near to the core.

The iron ore fields of Guyana form the **resource frontier**: a newly settled area in which resources have been recently discovered and exploited.

Case Study: Venezuela

Stage 1: Cities are few, scattered and small. Most roads run North to South.

Stage 2: Oil is discovered on and near the coastal strip.

Stage 3: Caracas grows as the single national capital.

Stage 4: Migration flows from the periphery (a backwash effect).

Stage 5: The core becomes attractive to investors and dominates the country while the periphery stagnates.

Stage 6: The core's popularity spills over into the periphery as oil development continues; sub and secondary cores are established.

Stage 7: Policy change to develop the periphery.

The Marshall reparation plans for MEDCs in Europe in the 1940s became the blueprint for LEDCs in the 1980s. However, pumping large sums of money into LEDCs has not helped their plight.

This is called a 'top-down' strategy.

Governments have largely accepted the concept of core v periphery. It is recognised that regional imbalances exist, and for a range of socio-economic reasons have to be addressed.

The cycle of poverty theory

This suggests that LEDCs will never reach the higher levels of development as they are too overly dependent upon MEDCs to do so. However, experience suggests that many of the large scale schemes, aimed at relieving problems in LEDCs, actually make problems worse, e.g. dam building in Brazil, Sobradenia has caused malaria to increase and any benefits that have accrued have been small and focused on the wrong groups of people.

Dependency theory

This suggests that the colonialistic tendencies of the past are perpetuated in the economic and political power that can today be exerted by MEDCs through agreements like GATT (the General Agreement on Tariffs and Trade) or through the edicts of the WTO (World Trade Organisation).

Only 7% of Europe's bananas come from the Caribbean.

The USA filed complaints, through GATT, against the EU schemes giving banana producers exclusive access to Europe's markets.

CASE STUDY

The secret life of a banana

If you want to see the face of unfair trade, go to rundown rotting Georgetown on the volcanic West Indian island of St Vincent. St Vincent's main export commodity is the banana. Windward Islands bananas are grown almost exclusively by peasant farmers. They get 10p a pound for top quality exports and just over half that for lesser quality bananas. Every 40lb box of fruit earns a farmer about £2 and costs another £3.50 to ship to Britain. By the time they get to the shelves the contents of the box may well sell for in excess of £50! Great profits are made by all except the grower.

But the economic injustice goes even deeper than the trading relationship between the vulnerable small growers and the powerful supermarkets. In the 1950s British foreign policy encouraged the West Indies and others to grow bananas and we established a tariff and quota system to insulate small growers against the Central American growers. The WTO has ruled it is 'unfair' to deny the market to all those that want to compete with the West Indians. What future now for the Caribbean growers of bananas and the exploited Central Americans? Both lose whilst we in the rich North gain.

Source: The Guardian November 1999

5.3 Causes of disparity

LEARNING SUMMARY

After studying this section you should be able to understand that:

- *the political and economic relationships between rich and poor countries reinforce the dependency of LEDCs on the MEDCs*
- *capitalism increases poverty and inequality in LEDCs*

Historic origins

AQA B	Some U4
EDEXCEL A	U5
EDEXCEL B	U5
OCR A	Some U5
WJEC	U4

In the 1960s the USA spent more than it earned → printed more $s → value of $ fell → oil is priced in $s therefore export value of oil dropped →1973 oil prices were hiked up considerably → large sums invested in world banks worldwide → interest rates plummeted as money was lent out too quickly → to re-coup losses the economies of the developing world are targetted → lavish amounts are lent, at

Comic Relief raised £26 million in 1997. This is paid back in debt repayments in just over one day!

Be aware how we fit into the indebtedness of the LEDCs!!

rates below inflation, and with no thought as to how it was to be used or repaid → some was used to repay other debts, about one-fifth went on arms, much was spent on development projects that proved of little value. By the mid 70s, encouraged to grow the same cash crops, the LEDCs found that they weren't getting the prices they expected for their exports → interest rates rose → oil increased in price again → the trap was sprung! LEDCs were earning less, paying more on loans and had to borrow more to pay off interest → effectively the LEDCs become bankrupt!

20% of Mexicans have no cash income. 30% earn less than $3/day.

Figure 5.9 *The indebtedness of the world's countries*

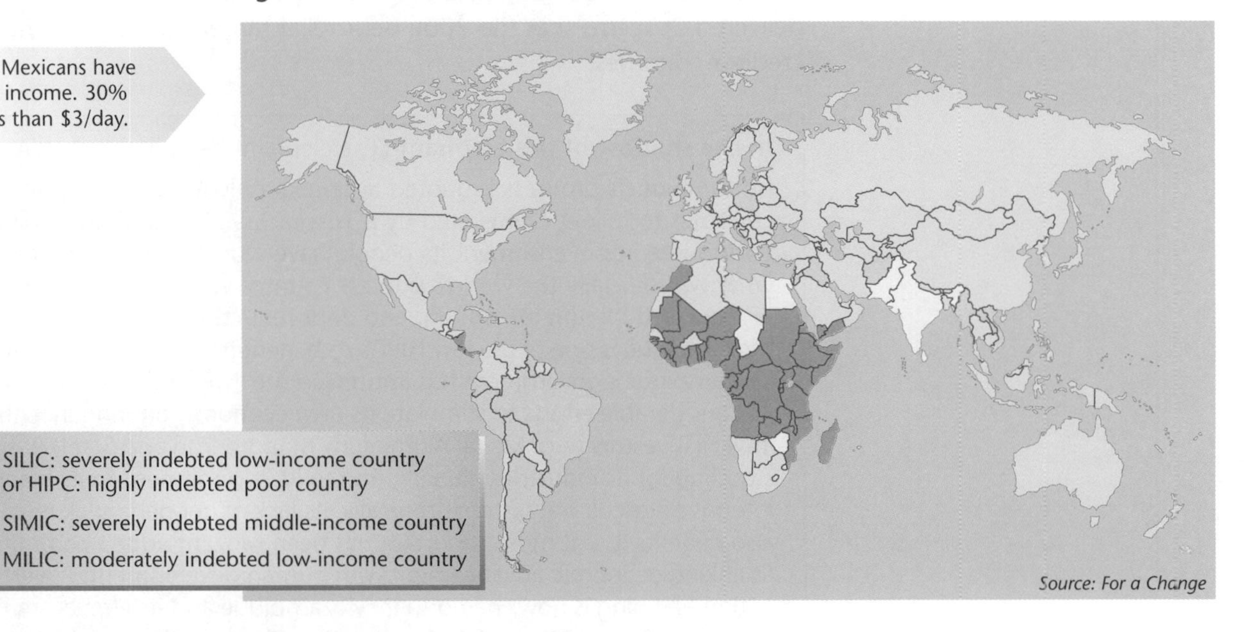

The principle cause of disparity: debt

AQA B	Some U4
EDEXCEL A	U5
EDEXCEL B	U5
OCR A	Some U5
WJEC	U4

Each year LEDCs pay the MEDCs nine times more in debt repayments than they receive in grants.

Each person in the Less Developed World owes £250 to the MEDCs!

Africa spends four times as much on debt repayments as on healthcare.

Knowing and understanding this sort of detail will put you at an advantage at A2.

Approximately 1.5 billion people in the world live in poverty. After many tens of years of economic growth and advance, large areas of the world have slid back into poverty. The enormous debts that have to be serviced affect millions of people. Latin America owes £365 billion and Africa £150 billion, equivalent to 83% of its GNP! Access to education is reduced: fees for education have been introduced in some countries causing a drop in enrolments. Healthcare is restricted: many LEDCs charge for healthcare, depriving the poorest who can't afford to pay! There is little employment; wages have dropped and employment levels risen and severely restricted trade; e.g. the IMF encouraged Mexico to grow cash crops, Mexico now imports many of the staple food crops it needs!

Responses to the debt crisis

The political response to the debt crisis has been varied and none of the initiatives have been adequate to deal with the whole debt crisis, see below:

The Brady Plan

According to this plan Brady suggests that banks should reduce the remaining debt of large debtor countries by writing off or re-scheduling their debt through conversion schemes. Usually converting it into sellable bonds!

Trinidad/Naples Terms

Initially a cancelling of half the debt of the poorest countries was suggested. Later a figure of about 70% was suggested and agreed at G7 summits.
This whole scheme hasn't worked, as the creditors have been reluctant to offer debt relief; and few countries would sign into the Structure Adjustment Programmes (SAPs) to reduce debt.

HIPC Initiative/Mauritius Mandate

(**HIPC** = Highly Indebted Poor Country)

This dates from 1996 and was a shift by the IMF and the World Bank to cancelling debts owed to them; to be financed by the sale of IMF gold, through a trust fund. HIPC initiatives have been limited in effect, Christian Aid reckons that only 6.4% of debt of the 41 poorest countries will be tackled. The Mauritius Mandate encouraged the key creditor countries of Japan, USA, Germany, France and the UK to put fresh impetus behind the HIPC initiatives.

To date only about £7 billion of the £70 billion promised has actually been delivered. It seems that the 2000 debt relief target hasn't been met by any of the creditor countries.

CASE STUDY

In the shadow of the elephant (The elephant being rich South Africa!)

Mozambique's progress is quoted as an example for all other poorer countries to follow. Devastated by a 16-year civil war and post-colonial Portuguese impoverishment its people have worked hard and reformed what was officially the world's poorest country. It is reaping the benefits of increased foreign investment and debt reduction.

Mozambique is one of the few HIPCs to benefit from the HIPC Initiatives; Mozambique's government has impressed the IMF amongst others with its considerable efforts to stimulate its own economy, offering incentives to inward investors (e.g. Mozal Aluminium/Maputo Iron and Steel). It has stood up for its industries globally, and to its commitment to its labour intensive agricultural system. Ironically, its lack of get rich quick resources also benefits it. But most important has been Mozambique's commitment to linking economic advancement with human development. Poverty Action Planning is now the norm in Mozambique and its effects are spreading through the whole country. Prioritised are the investments being made in education (in the 1970s as many as 90% were illiterate), health (Italy in particular has supported health), water and sanitation.

Source: Developments 1999

Barriers to tackling debt relief

- Poor countries have to prepare Action Plans or **SAPs** (**SAP = Special Adjustment Policy**), which have to be approved by the IMF and World Bank.
- Most HIPCs lack the skills to draw up such proposals.
- IMF and the World Bank demand that all areas of an HIPC economy and development are tackled.
- IMF and the World Bank are reluctant to write cheques for money that might be spent on weapons rather than adjustment programmes.
- Creditor governments and their opponents, in time-honoured fashion, rebuff and dispute the releasing of money to the IMF.
- Even if a country is offered relief it doesn't always help! Tanzania has had 20% of its debt written off. However the £59 million it still has to pay back is more than the annual spend on education. (A similar situation exists in Mali, Burkina Faso, Mozambique, Zambia and Malawi.)
- SAPs impose such onerous restraints on HIPCs that some have refused to co-operate. To meet IMF requirements countries may have to impose increased interest rates. Riots ensued in the Dominican Republic after food prices were doubled and medicine prices quadrupled (112 died and 500 were wounded in 4 days of rioting). In Argentina and Zambia strikes and demonstrations have been common; in the Venezuelan capital Caracas where wage levels collapsed and food subsides were cut, up to 1500 died in rioting in 1989.

Zambia's interest payments are at present $136 million; by 2002 they will be $235 million /yr!

SAPs mean spending less on health, education and social services and devaluing currency. Food subsidies are removed. Jobs go. Subsistence agriculture dominates.

The G8 Okinawa meeting in July 2000 actually backtracked on all previous agreements to relieve debt.

- The effect of the imposition of SAPs and the increasing unrest of individuals in the developed world with reference to the lack of progress with debt relief has lead to bloody confrontation in several developing world capitals, e.g. The Seattle Riots and riots in the financial heartland of London in the late 90s.

Consequences of disparity

AQA B	Some U4
EDEXCEL A	U5
EDEXCEL B	U5
OCR A	Some U5
WJEC	U4

The 'Boomerang' effect on the developed world

The debt of the HIPCs also affects the rich world.

The environment

Brazilians cut 50 000 km^2/yr of TRF to help service its US $112 billion debt.

The easy solution to the debt situation is for HIPCs to milk their own resources, by overfishing, by massive deforestation, by overusing soil to grow cash crops for the North. Plans set by SAPs to develop large dam projects and to drive industry with charcoal have also wrecked the environment. All of this has consequences for world climatology.

CASE STUDY

Made in Vietnam – cut in Cambodia

In the last 30 years Cambodia's forest cover has declined from over 70% to around 30% of land area. The forests have suffered an almost unprecedented assault from various warring factions and political parties seeking to fund their political and military aspirations. These illegal loggers rely on a ready market for their timber, and during the past four years a major section of this market has been the boom in the garden furniture trade. Garden centres and other retailers throughout Europe are seeking garden furniture made in Vietnam.

The trade in hardwood garden furniture is big business and is getting bigger. Imports of garden furniture into Norway in 1998, for example, were 95 times more, in monetary terms, than what they were in 1990. In the UK, a leading supplier has predicted that 1999 and 2000 will be boom years for garden furniture sales.

In many instances this furniture is marketed on the basis that it is environmentally friendly: 'For every fallen tree, a new one is planted so no tropical rain forest needed to be destroyed'. The reality of the situation is completely at odds with these claims.

By buying Vietnamese garden furniture consumers risk finding out that they are at best contributing to forest destruction in Vietnam, Malaysia, Myanmar (Burma) and Laos; countries that in part provide some of the timber used in the manufacture of the furniture. At worst there is a direct link between much of this garden furniture and the enriching of military warlords and the political elite in Cambodia.

Source: One World

Know your case studies.

Dole queues

In the MEDCs queues have lengthened as HIPCs under SAPs agreements have 'earned more, but spent less'. Before the debt crisis broke, Europe sold one-fifth of its exports to Africa. Through the 1990s this reduced to less than one-tenth.

The Drugs Trade

In an attempt to repay foreign debt HIPCs have increasingly raised currency by increasing the amounts of drugs, cocaine and cannabis, they grow, e.g. 41% of Bolivia's workforce use the drugs trade for their livelihood.

Of the 96% of export credits owed to the Department of Trade and Industry in the UK, 50% are owed from/to exporters for arms sales!

War

Debt leads to and contributes to war. The escalations of conflict in various countries leads to MEDCs intervening, e.g. the USA in Somalia, the British in Sierra Leone and so on.

The Banks

One of the winners in this entire debt problem has been the commercial banks. In Britain all the high street banks have lent to the HIPCs of the world. But through the bizarre system of debt exchange and tax relief in the so-called secondary market all have made substantial profits whilst debtor countries have gained nothing!

Effect on the less developed world

Effects of debt on children

There are thought to be 5 000 'street kids' in Guatemala City. So called 'street cleansing' of 'undesirables' accounts for the deaths of up to 300 children in Honduras, similar numbers in Brazil – usually at the hands of the security forces!

Families in HIPCs have to cope with a great deal. Lack of health provision impacts on family planning availability, resulting in many unwanted children. Parental inability to feed, clothe and provide for their children leads to many being abandoned onto the streets. In Central America this situation has become a massive problem for the authorities. Robbery, prostitution and begging are rife, all an essential source of income for the 'street children'!

Effects of debt on housing issues

The move to the cities also increases pollution in city areas.

Uneven development results in migration, within and between countries, usually from rural to urban areas. The cities have attracted many millions of these migrants, exacerbating housing problems. Many thousands of shanty towns (favelas, bustees, ranchos) have sprung up to accommodate growing populations. Seen as symbols of underdevelopment they have been demolished and bulldozed by HIPCs to enhance their SAP arrangements. Most of the shanty dwellers have been moved into housing for which they cannot afford to pay rent, and as a result these areas are themselves rapidly falling into decline, and so the cycle continues.

Effects of debt on women

In the world, women perform two-thirds of the work for only 10% of world income!

The active role of women in the development cycle was recognised in the 1970s. Developing countries now try in all SAPs to recognise and develop the skills and potential of women. The links between the formal education of females and their contribution to development and relief of debt is shown below.

Figure 5.10

The importance of female education

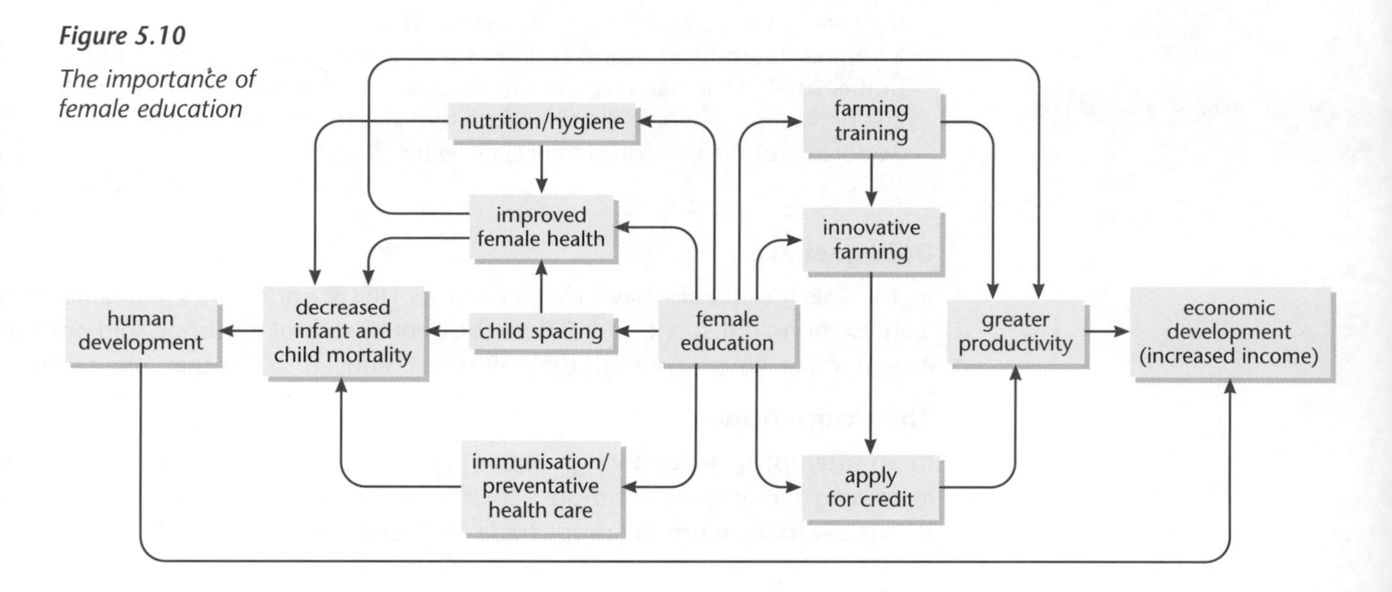

5.4 Aid and trade as forces for change

LEARNING SUMMARY

After studying this section you should be able to understand that:

- *aid may do more harm than good in the long term*
- *and competition can result in economic growth in LEDCs*

The burden of aid

AQA B	Some U4
EDEXCEL A	U5
EDEXCEL B	U5
OCR A	Some U5
WJEC	U4

Be able to show how we help LEDCs

'With at least four-fifths of the world's wealth the Northern MEDCs have a moral obligation to help the developing countries of the world', the Brandt Report.

KEY POINT

What is aid?

- **Short-term aid:** supplied after disasters, includes tents, medicine and food.
- **Long-term aid:** usually financial assistance or equipment, advisors and technicians.

Terminology:

- **Multi-lateral aid:** reaches the HIPC via organisations like the IMF.
- **Bi-lateral aid:** goes directly from the donor country to the developing nation.
- **NGO aid:** derived from NGOs (Non-Governmental Agencies) e.g. Oxfam.

Remember

- Aid tries to correct the imbalance in world resources and wealth.
- Poor countries still back their own development. MEDCs finance less than 20% of development projects undertaken.
- Profits in MEDCs is greater than the aid that is offered to LEDCs.
- Official aid sends millions more to LEDCs than the Non-Government Agencies (NGOs).
- The aid that is offered rarely reaches the people who really need it; some is used to pay for debt repayment, some is used to support the administration of aid. There is spending on weapons and prestige projects. In some cases less than 10% of the original aid reaches those that need it! In many cases smaller sustainable projects would be much more appropriate.
- Aid distribution is often tied to a particular government's pet political or economic agenda. It may also be tied to the purchase of donor country goods and services. Or to military access.
- Some people feel that aid actually nurtures dependency, and that trade should be encouraged rather than 'stop gap' aid being offered.

Aid-based development can cause 'aid fatigue' in donor countries.

CASE STUDY

Damning the dams

The building of dams is a hot political issue in HIPCs. Big dams are seen as misguided, inappropriate, technology-driven developments that have devastating effects on the environment and massive social consequences for indigenous people. A moratorium on World Bank loans for dam construction has been called for.

The UK's involvement in HIPC dam construction has also been halted after adverse reports in the British media linking £234 million of British aid for the Pergau dam in Kelatan Province in Malaysia, to the RM4 billion purchase of British fighter aircraft, this was in the period 1994 to 1996.

A consortium of MEDCs has withdrawn support for the Bakun Dam, in Malaysia. The future of the $5.5 bn project is now unclear.

N.B. Malaysia is in desperate need of clean, reliable water!

Trade and the growth of NICs

AQA B	Some U5
EDEXCEL A	U5
EDEXCEL B	U5
OCR A	Some U5
WJEC	U4

As mentioned previously many see the best route forward for HIPCs is to 'trade, rather than rely upon aid'. Since the 1970s some stronger economies have emerged from the developing world. These are linked either to oil discoveries or to the growth of NICs. NICs (**Newly Industrialised Countries**) have gained a significant amount of world trade over the last 30 years. Those countries that have been successful have on the whole protectable markets and produce a product that has enabled them to expand abroad.

Common characteristics of NIC development

- Poorly educated workforce
- Cheap labour
- They had early foreign assistance and investment
- They have co-operative governments who want to industrialise and develop
- There is lots of investment abroad, to broaden and develop export markets
- There is a strong work ethic, values and beliefs.

For South Korea, NICs have enabled export greater than imports. Their trading partners have increased and total corporate debt has dropped.

CASE STUDY

Strategy for the development of successful trade regimes and growth in HIPCs/LEDCs

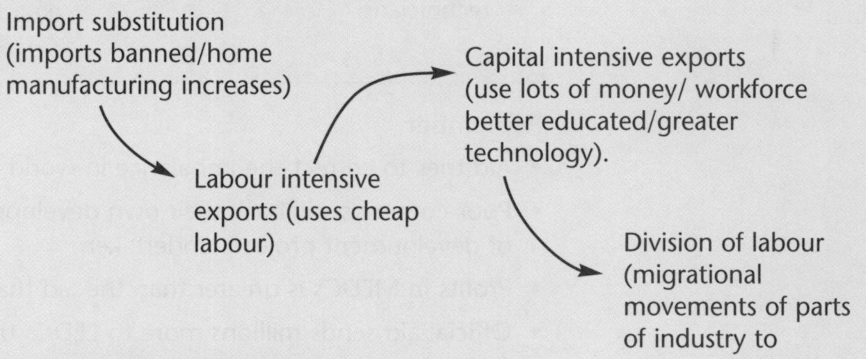

Import substitution (imports banned/home manufacturing increases)

Labour intensive exports (uses cheap labour)

Capital intensive exports (use lots of money/ workforce better educated/greater technology).

Division of labour (migrational movements of parts of industry to 'cheaper countries')

South Korean Development

The USA initially heavily supported the economy of South Korea, but the advantages of deep-water anchorage, markets on the Pacific Rim and a cheap hardworking labour force and government backing also helped. Much of the economy of South Korea is now in the hands of giant corporations (or chaebols) like Daewoo, Hyundai and Samsung. Initially the economy focused on textiles and heavy industry, it has shifted of late to high-technology goods. It has spread production abroad. 'Big players' also fall hard during periods of recession in world markets. South Korea was hit hard in the recent downturns in circum-Pacific markets.

Probably the strongest of the 'Asian Tigers'. By 1996 South Korea had in excess of 4.5 billion US$s invested abroad.

What future the NICs?

The global shift and globalisation of economies will lead to LEDCs being looked upon differently over the next century. As **GATT** (General Agreement on Tariffs and Trade) spreads the message of economic liberalisation it seems unlikely that LEDCs will be left behind economically. NICs will continue to prosper and develop, despite competition from TNCs (**Transnational Companies**).

Don't confuse NICs with TNCs in the exam room!

TNC = Transnational Company.

The threat of TNCs

AQA B	Some U5
EDEXCEL A	U5
EDEXCEL B	U5
OCR A	Some U5
WJEC	U4

All advantages and disadvantages need to be discussed in the exam room.

The TNCs and NICs have a huge influence over the world. This is a commonly examined area.

KEY POINT

The problems that TNCs will pose to the NICs in the future

TNCs operate and have ownership of assets in more than one country. They wield huge economic, political and financial power. Supporters of TNCs suggest they have fuelled development, critics suggest and focus on their moral and social effects.

TNCs exist to:

- gain access to new markets
- avoid trading barriers
- diversify
- reduce costs
- exploit lax environmental rulings
- provide new export routes.

Advantages of TNCs

To host:

- multiplier effect
- income generator
- introduction of technology
- employment growth

and home:

- profit.

The disadvantages of TNCs

To host:

- local companies lose out
- loss of autonomy
- spread of technology is suppressed
- lax safety regulations
- lax environmental stewardship
- pricing is manipulated to reduce tax liability
- financial costs to the host
- goods produced are inappropriate to the country

and home:

- possible job losses.

TNCs will be forced to take a responsible approach to world trade, so the NICs' future looks to be assured. Certainly the existence of TNCs look, on the whole, to benefit more than hinder HIPC/LEDC development.

'Alternative' strategies

AQA B	Some U4
EDEXCEL A	U5
EDEXCEL B	U5
OCR A	Some U5
WJEC	U4

'Bottom-up' policies in developing countries

These are smaller scale and more appropriate for many local and national needs. Most development strategies have a developed-world origin. As developing countries have different aims and structures, alternative strategies to enable development and sustainable development have to be instituted. Usually this involves appropriate and or intermediate technologies (see figure 5.11).

Figure 5.11 Advantages of using appropriate technology to meet local needs

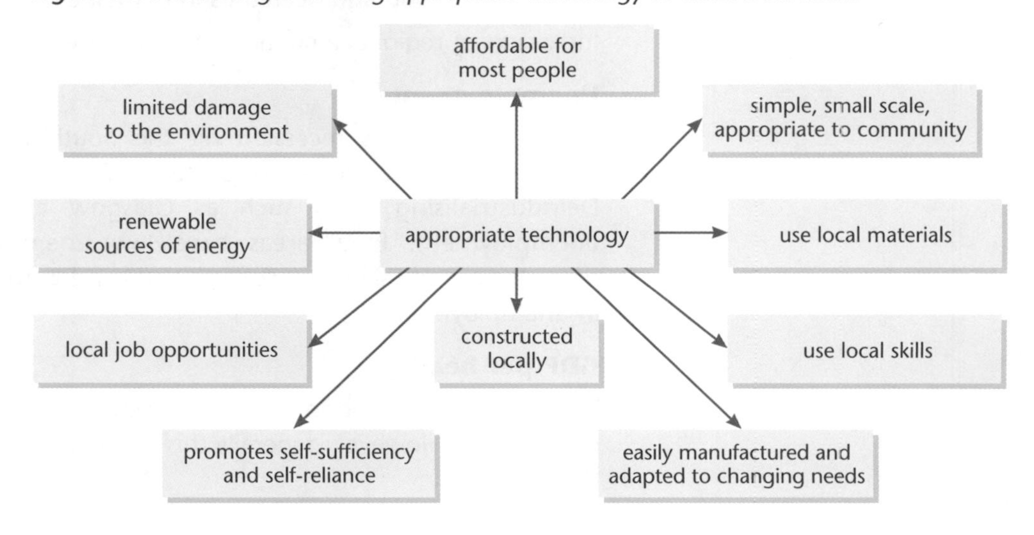

Teddy Exports of Tirumangalam, Southern India

Teddy Exports employs 300 local people and produces a range of products from massage rollers, cotton bags, shirts, lampstands and hair scrunchies. The business started in 1990 from a mud hut with five workers. Turnover is now about £1.5 million/year. Workers receive free medical care, subsidised food, housing loans and good wages. From the profits the Teddy Trust was formed, this has funded schools, evening classes, HIV/Aids awareness programmes, vet camps and a 'teaching' farm. Many of their products come into the MEDC marketplace, e.g. Bodyshop.

CASE STUDY

Agenda 21

In 1992 the United Nations conference on Environment and Development was held in Rio de Janeiro in Brazil. Usually referred to as the Rio Earth Summit, one of the major outcomes agreed by 180 of the world's leaders was a blueprint for sustainable development. This action plan, beginning in the 1990s but projected forward into the 21st century, is called Agenda 21. It looks at environmental, social and development issues and how they inter-relate. Formulating, agreeing and implementing strategies for sustainable development is now a requirement for national governments.

5.5 Inequalities in MEDCs

After studying this section you should be able to understand that:

- *every country, whether developed or developing, has regional variations in wealth*
- *regional development policies aim to spread wealth and development*

LEARNING SUMMARY

Inequalities in Britain

AQA B	Some U4
EDEXCEL A	U5
EDEXCEL B	U5
OCR A	Some U5
WJEC	U4

Realise that it is not just LEDCs that have disparity problems.

Thus far we have focused almost exclusively on the development issues that face HIPCs and the other major LEDCs of the world. Within countries of the MEDCs inequalities also exist.

In Britain the South is the richer area, containing the capital with its concentration of governmental and corporate power and a dense collection of rich towns. The towns on the whole have developed from the boom in producer services, business services and hi-tech industries. In the North cities are still trying to recover from deindustrialisation and decreases in manufacturing production.

Indicators of regional inequality in Britain are:

Unemployment

In the early 1990s recession hit the South harder than the North and the unemployment gap lessened. Since 1997 this gap has started to increase again. Deindustrialising cities such as Glasgow and Liverpool have the highest unemployment. Rural areas have low unemployment. With more children dropping out of school in Scotland and London this can lead to regional anomalies in unemployment.

GDP per head

This is highest in London, but is generally higher in the South and South East than in the North. However, generally this is a reflection of economic transactions rather than incomes.

Household tenure

There is no real difference in the number of owner-occupied homes between the North and South. However, homes in the South are more expensive (especially in London). But, London also has the highest levels of over-crowding.

Mortality

Life expectancy is highest in the South and lowest in the cities of the North.

Migration

People are migrating out of the North and London to the SE and SW.

Even these simple indicators show a clear North v South divide in the UK.

The case of Europe

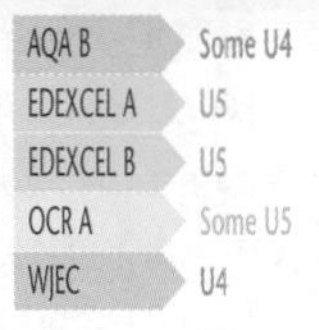

CASE STUDY

At a regional level – Core v Periphery Europe

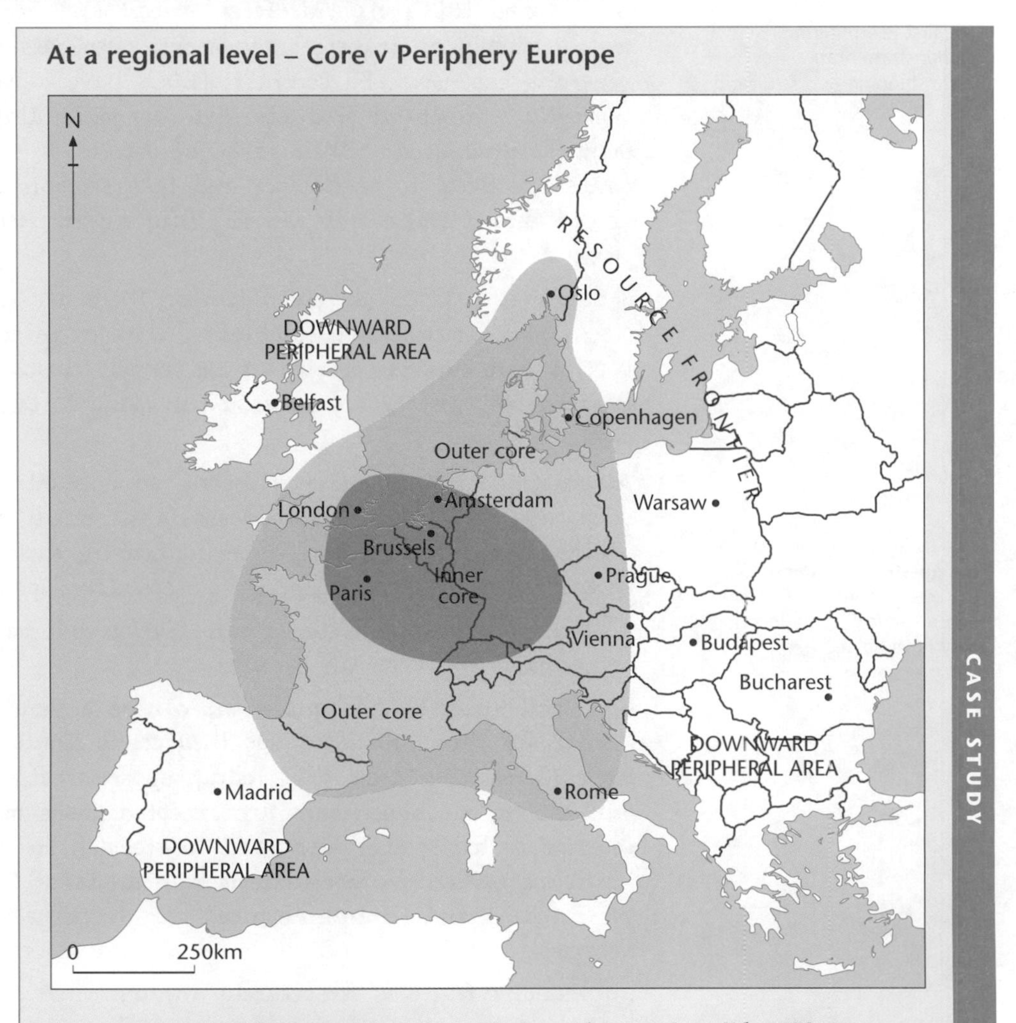

At a sub-regional level – France and the Mètropoles d'équilibres

The past Since recorded historical time Paris has dominated Northern central France, both socially and economically, the rest of France being effectively the periphery.

Decentralisation Encouragement of growth away from Paris has pre-occupied planners for the last 40 years. Initiatives set up under the Délégation de l'Aménagement du Territoire et à l'Action Régionale encouraged industrial growth, more extensive farming and tourism centres away from Paris, in growth poles in places such as Grenoble, Lyon, Marseilles, Bordeaux, Rennes and St Etienne.

Success Mixed, but the south is growing as a result of France's decentralisation schemes.

Sample question and model answer

1

Using examples discuss the ways in which trading has led to increasing economic dependency in LEDCs on MEDCs since the end of the colonial period.

This question is somewhat different from those normally set on the subject. It requires both an overview and evaluation of the dependency focus.

Trade is vital to the development of many LEDCs. Unlike aid which is mainly one-way, trade is a two-way process which can allow countries to develop economically. However, trade can also lead to underdevelopment. According to Franks Dependency Theory, LEDCs have not always been underdeveloped but they become underdeveloped or less developed because of trading relations with developed countries or empires.

The first few paragraphs demonstrate both understanding and familiarity with the topic.

MEDCs and LEDCs have different export and import patterns. MEDCs export mostly machinery, transport equipment, chemicals and services that are generally expensive. By contrast, LEDCs have a much smaller range of exports – mostly agricultural products and raw materials, which are cheaper than manufactured goods. Their range of imports is similar to MEDCs, although the goods are likely to be cheaper and less sophisticated. Hence, LEDCs are forced into a trade trap whereby they export cheap goods but import expensive products.

Trade in the 1990s is very dynamic. There are a number of new trading blocs, whose membership fluctuates, changes in tariff levels, restrictions on imports and exports, opening up of former 'closed markets' such as the Eastern Bloc and China, recession in many MEDCs and continuing debt problems.

This paragraph returns us to the question focus and sets the scene for the case study. Some 'tactical' underlining may have helped here?

According to the dependency theory, some countries become dependent upon stronger, frequently colonial powers, as a result of exploitation, trade and 'development'. As the more powerful country exploits the resources of its weaker colony, the colony becomes dependent upon the stronger power.

To illustrate how dependency can develop and be exploited I will be looking at the banana trade as an example.

For small-scale banana growers it can be a very tough life, and it is getting harder for them. 10 years ago they could simply cut the bananas, pack them loosely and send them off. Today, supermarkets will reject the tiniest bruises. No misshapen ones are accepted, none must be too long or too short, too ripe or too spotty. Also hard to stomach for the farmers is that a hurricane (which are not uncommon in areas of banana growing, such as the West Indies and Central America) can wreck years of work in just a few seconds.

Good use of a specified example. To cover the development topic effectively, you have to have up-to-date examples in your armoury!

The banana trade is desperately unfair. Windward Island bananas are grown almost exclusively by peasant farmers. They get 10p a pound for top quality exports and just over half that for lesser quality. Even the top price barely covers the cost of insecticides, fungicides, fertilisers and the labour needed to produce the fruit that the supermarkets require.

Every 40lb box of prime fruit earns a farmer about £2 and costs another £3.50 to ship to Britain. By the time they get to the shelves – ripened to a shiny, uniform yellow, pretty as sweets, each a regulation size, shape, taste, weight and price – the contents of that box may sell for nearly £50.

The economic injustice is far deeper than the trading relationship between the vulnerable small growers and the powerful supermarkets. Ever since the 1950s, British foreign policy encouraged the West Indies and other commonwealth

Sample questions and model answer (continued)

countries to grow the world's favourite fruit and established a tariff and quota system to insulate small growers from booming central American plantations, the islands have depended on us. This is a prime example of people in LEDCs being heavily reliant on MEDCs.

The situation has now been made worse with the WTO ruling that it was unfair to deny a market to the other giant agri-business corporations and USA. There is no way farmers on hillside patches will ever compete on price in a free market with 10 000-acre industrialised plantations.

Notice how this student is able to keep the 'thrust' of the essay going right to the bitter end!

Many people believe the only fair system is the 'fair trade' model that has evolved. All products are guaranteed by a mark, which is issued after inspection and certification. Currently, when we buy a cheap banana, we are unwittingly participating in exploitation. With fair trade we would be paying more for our bananas but we should remember there are mothers, fathers, children, blood, sweat and toil behind the produce and we are only being asked to pay what it costs to produce.

Comment:

It is imperative that you can write on topics like the above using contemporary and up-to-date information. This is an excellent essay, it is coherent, maintains the reader's interest, but importantly answers the question! The author of the essay gained an A in his terminal exam and is now studying the subject at university.

Practice examination questions

1

A Study the map which shows a classification of the level of development for countries of the world.

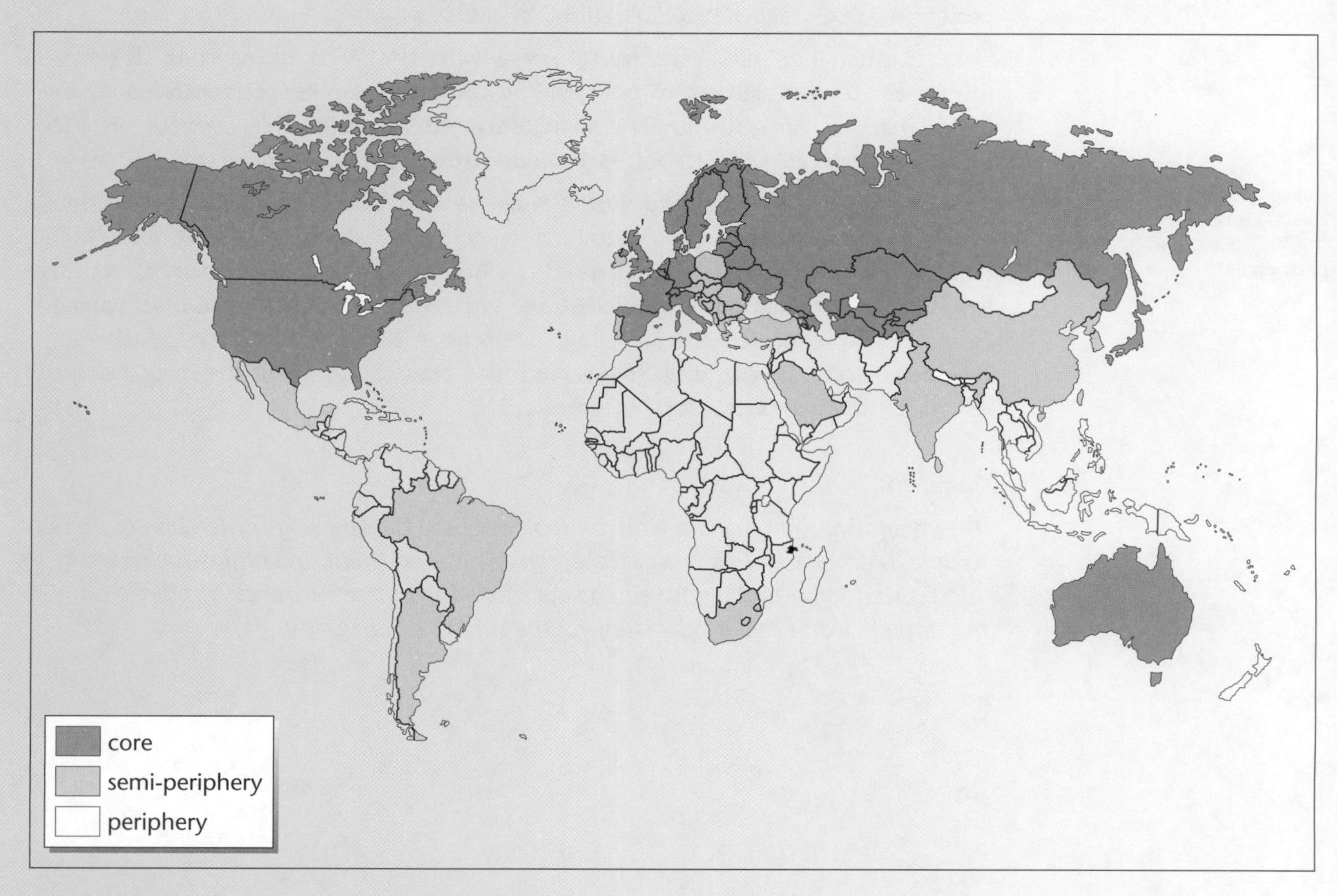

(a) (i) Define the terms:

1 Core;

2 Periphery. [2]

(ii) Describe the distribution of:

1 Core areas;

2 Semi-periphery areas. [4]

(iii) Suggest the criteria that could be used to classify countries as semi-periphery. [4]

(iv) Explain the distribution of the semi-periphery areas. [5]

(b) Outline how physical factors might affect the development of the periphery. [5]

B Explain why and how government policies can encourage regional economic development. [20]

EDEXCEL 1998 (6204 9202/4)

Practice examination questions (continued)

2

Study the diagram below taken from a bill-board photographed in Accra, Ghana.

(a) Outline the possible advantages to LEDCs of exporting primary products. [5]

(b) Examine the role of trade and aid in the growth and development of the world economy. [20]

EDEXCEL specimen paper for new AS/A2 exams

Chapter 6

Leisure and tourism

The following topics are covered in this chapter:

- *The theoretical framework*
- *Trends in tourism in the UK*
- *International tourism*
- *Sustainable ecotourism*

6.1 The theoretical framework

LEARNING SUMMARY

After studying this section you should understand that:

- *the term leisure includes recreation, sport and tourism*
- *leisure participation is influenced by economic, social, cultural and political factors*
- *leisure exists on a demand supply circle and makes demands on resources*
- *the demand for tourism is constantly growing*
- *tourism can be classified using a range of terminology*
- *growth, stagnation and decline of tourism can be modelled*
- *the features and qualities of tourists can be modelled*

The leisure industry

AQA B	U5
EDEXCEL A	Some U5
EDEXCEL B	U5
OCR A	U4
OCR B	U5

Leisure can be defined as 'time away from work, paid or unpaid (housewives); time when you are free from other obligations and where you choose to be involved in an activity very different from the usual things you do'. Like all human activities leisure impacts on our resource base, both locally and globally, and our environment. Leisure is demand driven and as such others profit by supplying various 'needs' (see figure 6.1).

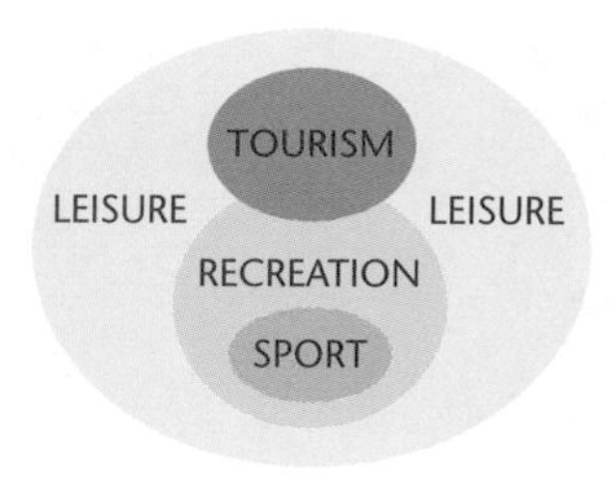

With more leisure and recreation time available it is important you know and understand its dynamics.

Figure 6.1 *Leisure: supply and demand*

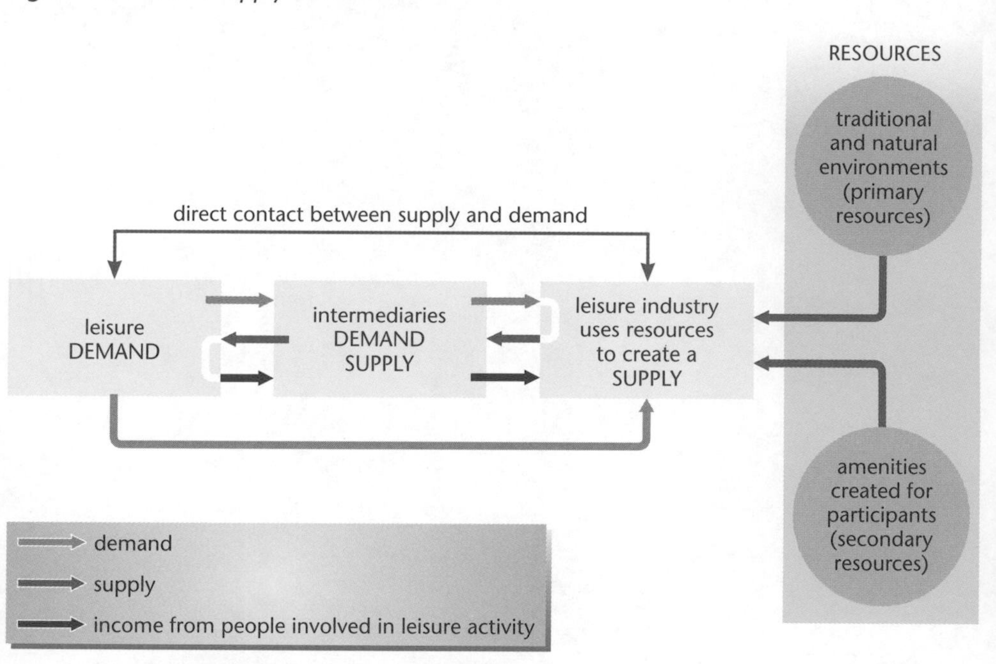

Key points from AS

- **The challenge of the coast**
  Revise AS pages 34–45
- **The challenge of the atmosphere**
  Revise AS pages 46–60
- **Worldwide industrial change**
  Revise AS pages 126–138

The economic potential of leisure is enormous, we spend up to 20% of total income on leisure. The leisure industry is now the biggest world business. Globally the industry is larger than the world trade in many raw materials e.g. iron and steel etc. It employs nearly 10% of people who are in employment.

Generating 10% of the world's GNP.

> **Britain's spending on leisure**
> As we enter the new millennium we are now spending more on leisure than on housing and food. One-sixth of household expenditure goes on leisure, that is 17% of income. The age group 65–74 spends the most on leisure and services.
>
> CASE STUDY

High Lodge, part of Thetford Forest; has tens of miles of bike trails and has been successful in attracting a large number of mountain bikers.

Leisure is socially and culturally linked, it is also influenced by fashions and fads, e.g. mountain biking in country parks. However, the basic factor underlying our leisure involvement is **time**.

Leisure can be defined in time terms, is voluntary and can be enjoyed at home or close to home. Tourism involves staying away from home for at least one night or more.

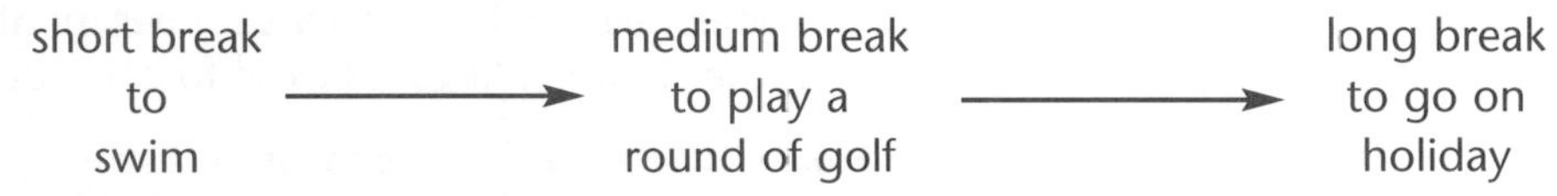

On the whole, people with limited leisure time tend to seek out and be involved in activities close to home, whilst those with more time travel much further afield (see figure 6.2).

The leisure/tourist continuum

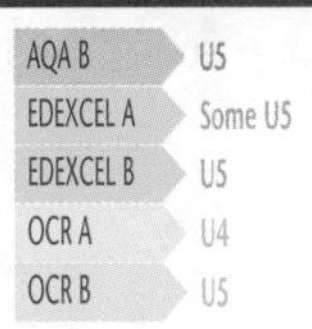

Figure 6.2 *The leisure/tourist continuum*

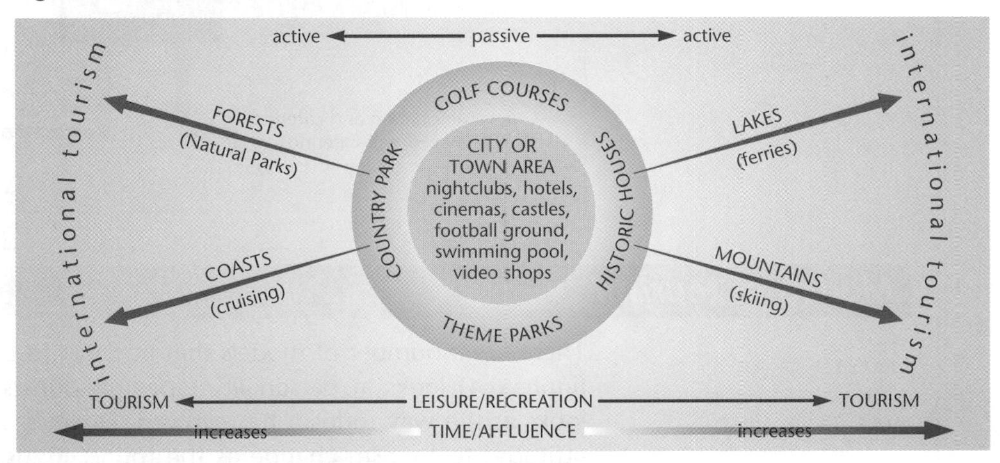

Analysing tourism

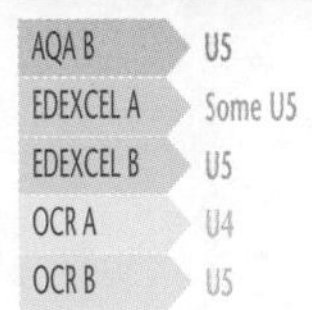

We all travel; know how it is classified.

> **Classification**
> - Domestic tourism – visits within own country
> - Inbound tourism – visitors to UK
> - Outbound tourism – destinations abroad
>
> **Why travel?**
> - Touring and sightseeing culture
> - To relax
> - To be involved in recreational activity
> - To visit friends and relatives
> - Business trips
> - Specialist trips
>
> **Where do tourists go?**
> Destinations that:
> - are attractive, e.g. beaches, mountains
> - offer the right facilities (amenities) i.e. hotels
> - are accessible, e.g. by air or easily by car.
>
> **How do tourists get to their destinations?**
> This is changing all the time. In the last 30 years there has been a 24% increase in air travel, a 16% drop in boat traffic. Transport is 'evolutionary', methods change, e.g. train to boat to plane.
>
> KEY POINT

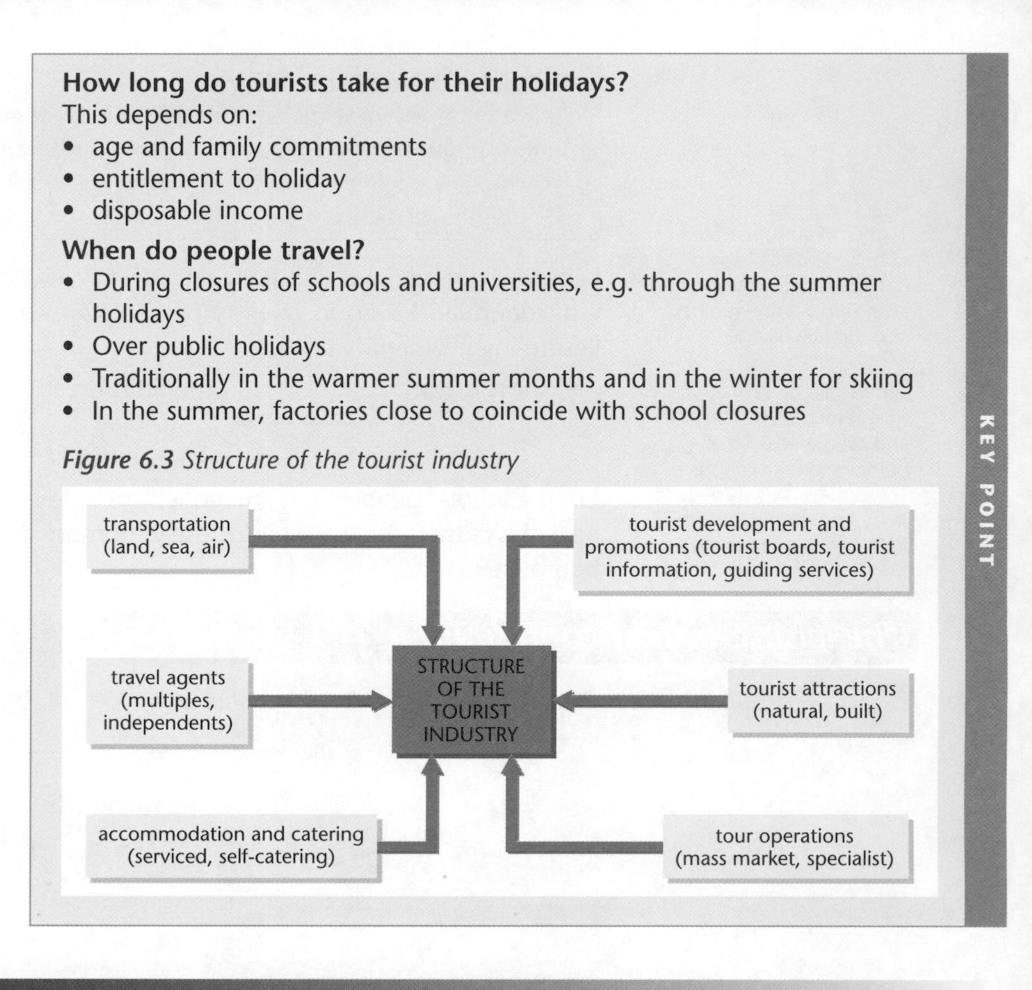

How long do tourists take for their holidays?

This depends on:

- age and family commitments
- entitlement to holiday
- disposable income

When do people travel?

- During closures of schools and universities, e.g. through the summer holidays
- Over public holidays
- Traditionally in the warmer summer months and in the winter for skiing
- In the summer, factories close to coincide with school closures

Figure 6.3 *Structure of the tourist industry*

Models of tourism

AQA B U5
EDEXCEL B Some U5
OCR A U4
OCR B U5

There are a number of models that attempt to classify tourism. Plog's model (see figure 6.4) looks at personality types of tourists; Butler's model (see figure 6.5) looks at the way tourism has evolved. Doxey's Index (see figure 6.6) shows how attitudes to tourism change as the tourist industry develops. The Enclave model (see figure 6.7) shows how socio-economic factors influence tourism.

Figure 6.4 *Plog's personality model*

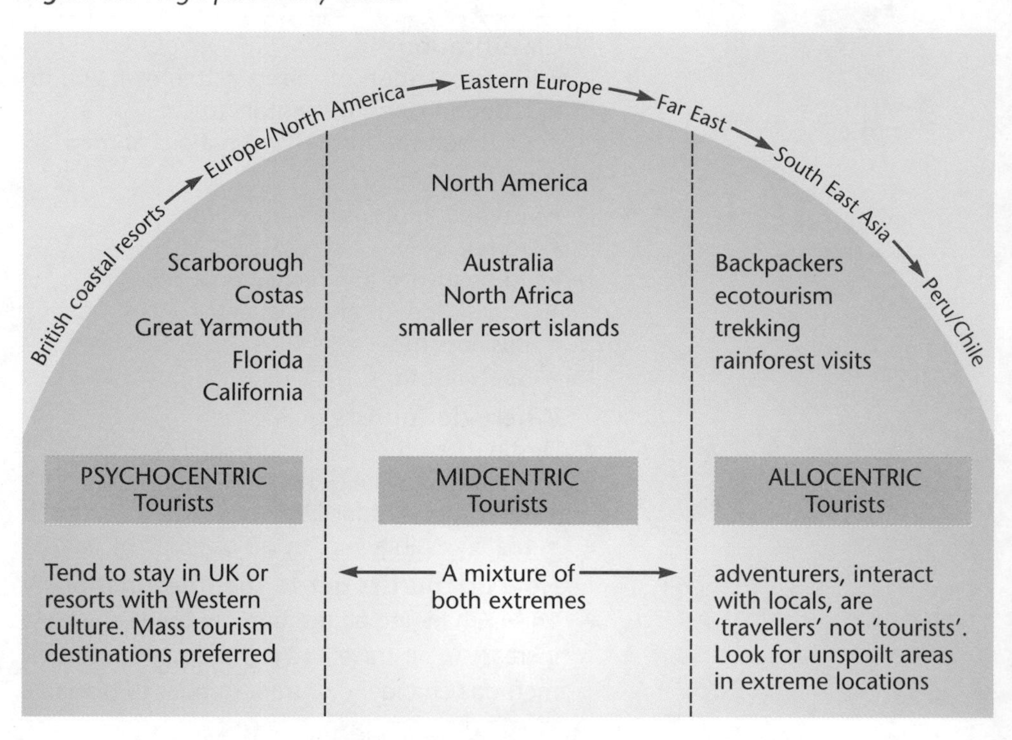

Figure 6.5 *Butler's tourist area life cycle, 1980*

Source: Edmunds At Leisure

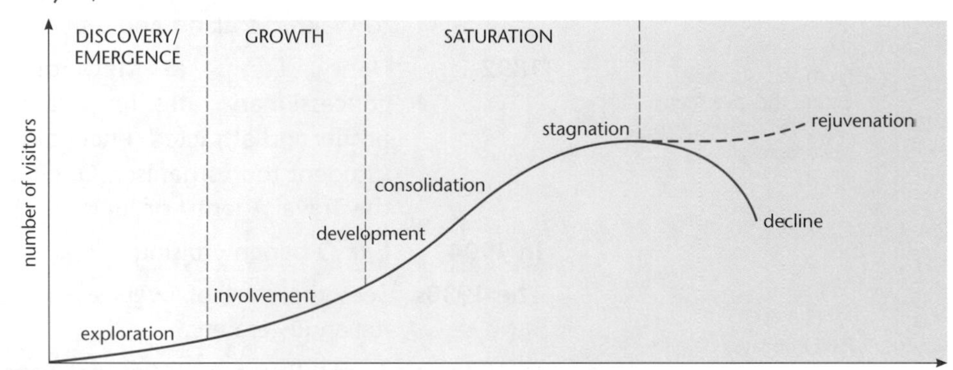

Exploration
- small numbers of allocentrics or explorers
- little or no tourist infrastructure
- natural or cultural attractions

Involvement
- local investment in tourism
- pronounced tourist season
- destination advertised
- emerging market area
- public investment in infrastructure

Development
- rapid growth in visitation
- visitors outnumber residents
- well-defined market area
- heavy advertising
- external investment leads to loss of local control
- artificial attractions emerge to replace natural or cultural
- mid centrics replace explorers and allocentrics

Consolidation
- slowing growth rates
- extensive advertising to overcome seasonality and develop new markets
- psychocentrics attracted
- residents appreciate the importance of tourism

Stagnation
- peak visitor numbers reached
- capacity limits reached
- resort image divorced from the environment
- area no longer fashionable
- heavy reliance on repeat trade
- low occupancy rates
- frequent ownership changes
- development peripheral to original developments

Decline
- spatial and numerical decrease in markets
- a move out of tourism
- local investment might replace abandonment by outsiders
- tourism infrastructure is run-down and might be replaced by other users

Rejuvenation
- completely new attractions replace original lures, or new natural resources used

It is possible to model tourism. This allows more accurate descriptive and analytical work. This sort of information can be incorporated into essays.

Figure 6.6 *Doxey's Index e.g. Bali*

Figure 6.7 *Enclave model*

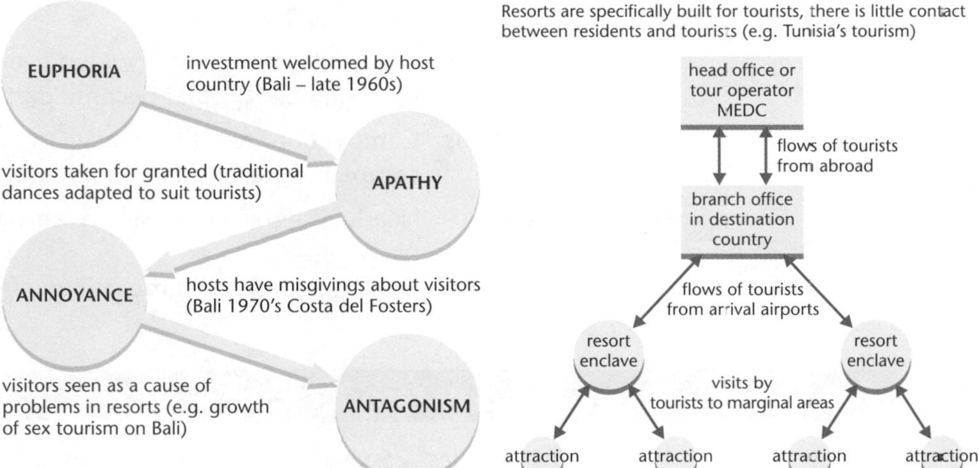

6.2 Trends in tourism in the UK

LEARNING SUMMARY

After studying this section you should be able to understand that:

- *UK tourism has changed greatly, with domestic holidays giving way to holidays abroad*
- *tourism is the UK's most important industry*
- *resorts have a distinctive character in the UK and have been forced to 'change' to survive*
- *the tourism product in the UK is diverse*
- *foreign visitors are important to the UK tourist industry*

The advent of the package holiday

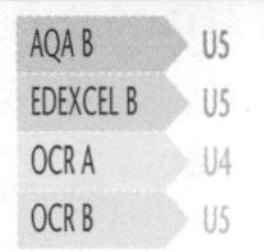

This section explains how tourism has changed. Change is a favourite with examiners!

1841 Thomas Cook in an attempt to wean people off alcohol arranged an excursion from Leicester to Loughborough.

By 1855 Thomas Cook was taking visitors to Calais.

Greater affluence.

By 1863 Thomas Cook expanded his undertakings, taking the working classes to Wales, Scotland and a first distant overseas trip to Switzerland.

From this discovery comes the 'pile them high, sell them cheap philosophy'.

1892 Henry Lunn, a Methodist ex-missionary, discovered that concessionary rates for guaranteed mass bookings, created large profits and attracted a new bigger group of tourists.
(Student tour organiser, Quentin Hogg, later joined Lunn to form the Travel Agents' group Lunn Poly.)

In 1904 P & O began cruising.

More leisure time.
Paid holidays.
Accessibility and transport improves.

The 1930s Sees the start of overseas holidaying for the rich; in Nice and the Italian Riviera etc.

In 1948 The **Holidays with Pay Act** gave 25 million UK workers 2 weeks paid break per annum.

In 1950 The first holiday company emerged. *Horizon* was established by Vladimir Raitz, offering package holidays and charter flights.

A Balearic holiday in 1957 would have cost £35 all in!

By 1955 Two and a half million holidayed abroad, at least one-quarter travelling by air.

Cruising peaked in 1957.

Greater mobility.

In 1968 Thomson became the first operator to use jet planes; this made journeys to resorts shorter for the holidaying public and cheaper for the travel firms.

Product development, including the lucrative pensioner market.

The 1970s The Mediterranean was within everyone's reach.
Package holidays had become part of our way of life.

The 1980s Competition for the market share of tourists by the big tour companies has ensured continuing low prices for holiday-makers.

The word tourism comes from the word 'autour' meaning around.

In 1999 Nearly 30 million travelled abroad, 15 million on fully inclusive packages.
Once the preserve of the rich and famous, the world is now accessible to all.

The traditional UK holiday resort

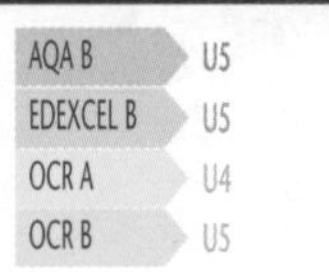

Over the last 25 years we have increasingly spent less on UK holidays. At present we spend nearly £25 billion on foreign holidays and about £17 billion on UK holidays. Clearly international tourism accounts for much of this change, see later section.

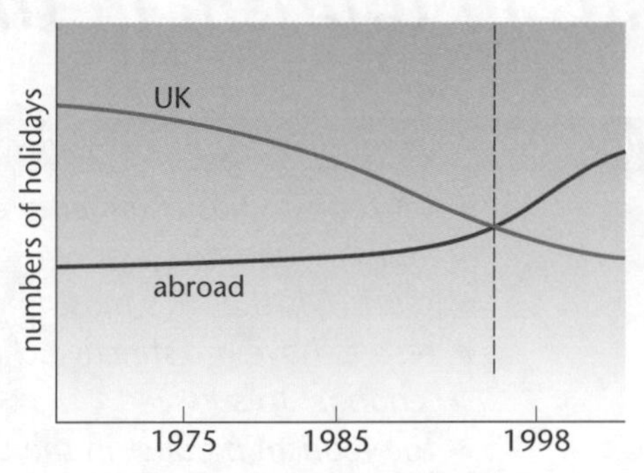

However, many UK seaside resorts still provide for domestic holidays. Each resort has its own unique location, morphology and degree of dependence upon its resort function.

Tourism has been described as the industry without chimneys.

Resorts are a distinctive type of urban settlement in the UK with characteristic land use patterns, explainable using bid-rent theory. (In a resort highest land values are found near the beach/sea with a peak where the frontage meets commercial property).

Figure 6.8 *Traditional seaside resort and pressures upon it today*

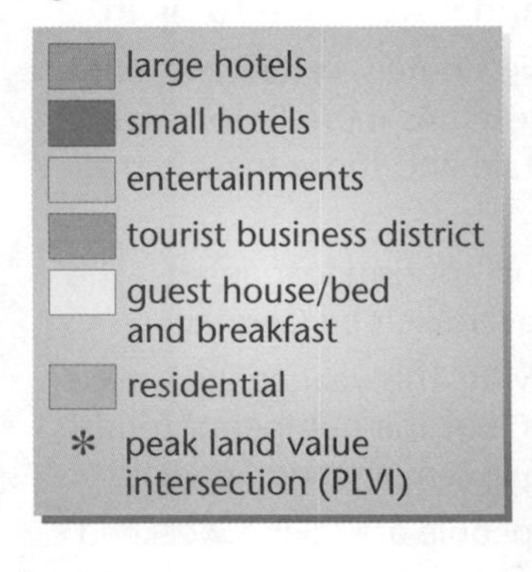

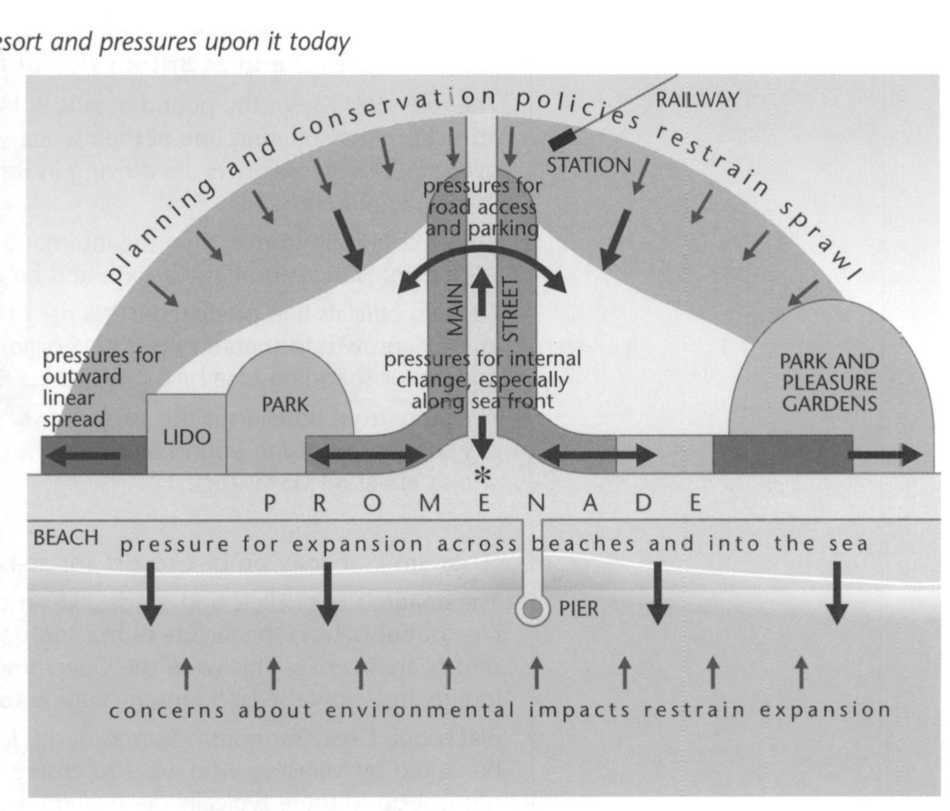

Source: Prosser, Leisure, Recreation and Tourism

Case studies are vital in your A2 studies.

Average temperature in summer = 16°C, with 619 mm of rain/yr.

67% of visitors arrive by car. 50% visit the resort year on year.

Conference facilities have been built, and a 10-screen cinema, holiday village and ski-slope.

CASE STUDY

'Bucket and spade' Scarborough

Scarborough, accessible by the A64, is to be found on the East Coast of Yorkshire. It claims to be the first seaside spa resort. It was fashionable for the Victorian aristocracy to use its excellent sea bathing facilities. The railway brought the middle classes to the town in 1846 and by the late 19th century Scarborough was well established with people from North and North East Britain. The morphology of Scarborough is similar to the model outlined above.

Over the last few years Scarborough has seen a switch from high-season tourism to year round visits. Tourism impacts on Scarborough in all the usual ways. On the positive side it generates massive amounts of income (in 1998, £270 000 000) and nearly 20% of all employment. The permanent population also enjoys many modern facilities. On the negative side, congestion, housing and pollution problems are all too common.

For the future, there are challenges to face. But strategic planning should ensure (with EU backing) that the town is able to compete. Recent developments have tagged Scarborough the 'National Seaside Resort of the 21st century'!

Scarborough continues to be successful, along with Great Yarmouth, Blackpool and Brighton. Its large resort status ensures it is unlikely to lose out on tourist numbers and decline. Many smaller resorts have declined, as reduced numbers of tourists spend less time in such resorts, and spent less money. To attract visitors and investment to survive, resorts have to have quality attractions, character and quality accommodation: this is known as **critical mass** in the tourist industry. UK tourism is susceptible to fluctuations in our economic conditions and to ever changing 'tastes'.

CASE STUDY

'Tourism faces slump as Britons fly out and visitors fail to fly in'

The relentless rise of the pound means Britain's hotels, restaurants and attractions are suffering one of their worst years. As more Britons head overseas, fewer foreigners are arriving in Britain and those that do make it are spending less.

Britain could slip from fifth in the international tourism league after the US, Spain, France and Italy to sixth and be overtaken by Germany.

Tourism officials had predicted a 3% rise in visits this year to 26.5 million and 5% growth in spending to £13.6 billion. But the number of tourists and visitor spending rose by a meagre 1.5% in January and February.

The story from hoteliers is the same. Fewer people are taking weekend city breaks. The rising pound and slumping Euro have had a particularly strong effect on US visitors.

Source: Observer, 7 May 2000

'Tourism tsar may be last resort for seaside towns'!

The image is of buckets and spades, kiss-me-quick hats and melting ice-cream. But behind the façade of bracing good humour Britain's seaside resorts are in crisis. This week the Government is to launch a major inquiry into how the UK's famous seaside towns can be saved.

Blackpool, Great Yarmouth, Scarborough, Minehead and Newquay will be visited by Ministers who want to ensure that regeneration projects do not just go to more typically depressed areas such as inner cities and former steel towns.

The English Tourist Council will investigate what people require from a British seaside holiday, and a task force has now been set up under Peter Moore, managing director of Center Parcs. Moore, already being described as the 'tourism tsar', will be expected to come up with new initiatives to save the seaside resort.

The British seaside tourist industry is worth nearly £5 billion a year, with 26.5 million seaside holidays taken in Britain last year. But the big figures mask economic depression and unemployment, particularly in winter. Seaside resorts generally suffer from higher levels of unemployment than the national average and lower rates of economic activity. High numbers of older people, many with little disposable income, mean that shops and businesses struggle to survive. They are also on the edge of transport networks, meaning that industries outside the tourism sector do not want to move to seaside towns.

Cut-price foreign travel has hit our resorts hard. Between 1965 and 1999 the number of Britons travelling abroad on holiday has risen from 5 million to 27.3 million.

The economic, social and environmental pressures faced by the seaside resort have at last attracted attention from Government.

Source: Observer, 30 July 2000

New initiatives

AQA B	U5
EDEXCEL A	U5
OCR A	U4
OCR B	U5

The UK tourist is becoming footloose with a greater, broader range of holiday opportunities on offer, to reflect both the UK's increased mobility and wide ranging interests. A number of examples are offered here.

Holiday villages

The idea of holiday villages is not new. Early holiday villages were built in lines of similar communal establishments. The first in the UK, a 'tented' community, was set up at the turn of the last century at Caister-on-Sea. The first Butlins Holiday

Camp was constructed at Barry, South Wales. Accommodation is cheap and cheerful and everything is provided on site for the tourist. In reality the village is a planned resort, or tourist enclave. Such villages probably originated in Continental Europe. In March 1985 a Dutch company brought to the UK a new type of holiday village. Opening its doors in 1987, Centre Parc now has three such centres in the UK. These parks rely upon a mobile public wanting to be involved in leisure activities in an all-year-round environment.

Heritage tourism

Often termed 'retro-mania'.

Nearly 55% of all domestic holidays now make use of the countryside and heritage attractions. All city and county areas have to have structure plans with tourism as a prime concern. Heritage in its broadest context allows visitors to look at nostalgia, tradition and historical customs, but is not without some problems; see below:

CASE STUDY

'Tourists threaten to swamp historic cities'

England's historic towns are in danger of being swamped by tourists. Popular destinations must not risk becoming victims of their own success.

A report, *Making the Connections*, was commissioned by the English Tourism Council, the English Historic Towns Forum and English Heritage. It is aimed at preventing the tourist boom from alienating town-dwellers and making foreign visitors feel like unwelcome, if necessary, contributors to the local economy.

About half of Britain's 26 million tourists from overseas are thought to visit historic towns, and nearly a third say these were an important factor in their decision to come.

The report encourages towns to pursue ideas such as the bike parks in Taunton, free shuttle buses in Cambridge, 'pub watch' scheme in Lincoln and 'residents first' weekend in York, when local people get free entrance to city attractions as a thank-you from the local tourist industry.

The new sustainable tourism must not harm the environment, must be acceptable to the community and be profitable for businesses – but also satisfy visitors, who are sophisticated, well-travelled and want to experience history and ambience in a way that cannot be achieved at a purpose-built facility such as a heritage centre.

Theme parks

Most theme parks are privately owned and determine their own admission charges.

Adapting and changing again. Know why this has happened.

It is said 'a successful theme park works by closing off the real world'. The British theme park is vastly different to its American counterpart; though thrill rides are in evidence there is much more emphasis on the educational aspects of the theme parks. So in the UK, theme parks range from the purely educational Iron Bridge Gorge Centre to the thrill-seekers paradise at Alton Towers. One feature of all these theme parks is that people are willing and prepared to travel long distances to reach them. The parks are highly accessible and are within easy reach of the densely populated parts of the country. In what is a very volatile market continual adaptive change is a must as visitors' tastes change. Windsor Safari Park is a good example, it was a leisure casualty and no longer exists. Legoland took its place!

Other attractions

These include farm visits, zoo attractions and animal sanctuaries.

Sport and leisure

As disposable income has increased so has peoples involvement in medium-to-high cost sports. This has led to a plethora of developments related to water-based activity. Demand has grown by 50–60% through the 1990s, for sailing, water-skiing, windsurfing, rowing and canoeing. Second generation sports centres have also started to appear in the urban landscape. These are a level above the 'original' sports hall, offering swimming, courts, fitness suites, climbing walls and other activities. They reflect both changing tastes and our increasing affluence; and the fact we have more time available for such activities.

Foreign tourists in the UK

AQA B U5
EDEXCEL U5
OCR A U4
OCR B U5

Some 32 000 000 tourists are expected to arrive in the UK by 2004. Most are from Europe, though US citizens continue to make up a significant proportion of our visitors. Annually, foreign tourists to the UK spend some £14 billion. Nearly 50% of all visits are focused on and within Greater London. The other half of our visitors congregate around tourist honeypots like Stratford-on-Avon, Oxford, Warwick etc.

6.3 International tourism

After studying this section you should be able to understand:

- *the nature of international tourism and the variety of resource it utilises*
- *that tourism conflicts with other land uses and ways of life*
- *that management strategies integrating values and interests are the way forward*

LEARNING SUMMARY

Dynamics of tourism

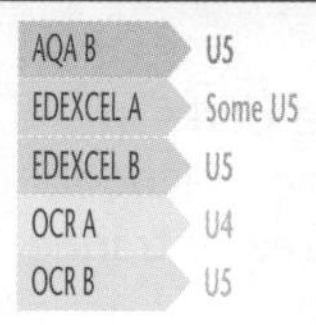

Choice of destination

Tourism is not evenly dispersed throughout the world, both physical and human factors affect the pattern of distribution.

Significant factors include:

- resource distribution (both cultural and natural)
- a country's ability to provide the infrastructure for growth
- weather and climate, temperature and seasonal variation of weather have an effect, and hotter areas in particular are advantaged!
- government controls – some allow tourism to grow in an ad hoc way, others regulate and plan, e.g. Kenya
- changing tastes
- war and 'political' problems can deter visitors, e.g. Yugoslavia and South Africa
- the state of the economy, business tourism is an important facet of tourism.

Opportunities to travel have increased. It is important you know why and what effect we have on the environment.

Figure 6.9 *Tourist destinations*

The four boxes on the right show the two types of country that the tourist travels from (**A** and **B**) and the two types he can visit (**C** and **D**)

There are four possible routes (i–iv). Developing countries hope that route (i) is where most growth will take place

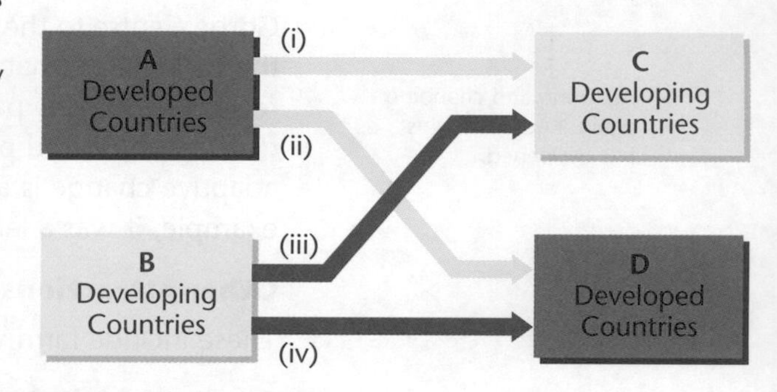

Reasons for growth

Despite the unevenness of its distribution, international tourism has grown because:

- by raising awareness of other places, education and the media have created a

demand on providers to allow us entry. Countries have become consumer packages
- the advent of the tourist industry has revolutionised the way we purchase holidays
- cheap charter travel, by plane and boat, makes travel possible for a wider audience
- relative affluence: increased standards of living, paid holiday and early retirements give people more time to travel.

KEY POINT

Are tourists killing the paradises they visit?

1.6 billion tourists currently roam the world (spending $2 000 billion): thus it is difficult to disagree with the need to minimise the impact of tourism.

Supporters of green tourism cite the ugly developments in Spain, building sites on the Turkish coast and the pollution of the sea in Thailand and the Philippines. The irony is that green tourists go to some of the most sensitive sites on Earth where environmental impacts may be just as severe. How many people can visit the Galapagos without affecting the ecological balance?

Tooth decay in children in Nepal increased dramatically a decade ago – and the cause? Climbers on the Annapurna Trail handing out sweets.

Then, there are the cultural imperialists who highlight the social changes that come in the wake of tourism – the damage to communities and local crafts.

Examples of damage left in the wake of tourists include:

- The anchor on a cruise ship weighs 5 tonnes, one swaying anchor can cause damage to reefs that takes 50 years to repair.
- Boat propellers leave scars on the ocean bottom. The silt stirred up causes sea-grasses to die and scares animals away.
- Manatees, a mammal that lives in Florida's rivers, are being mangled, crushed, trapped and killed by power boat propellers, canal locks and discarded fishing gear.
- Despite objections from the aboriginal Ngarrindjeri people, tourists continue to travel from Goolura to view aboriginal remains near Marks Point in the Coorong National Park. For aboriginals these remains are sacred and to profit from them is profane.
- China's 2000-year-old terracotta army, near the northern city of Xian, is under threat; and tourists' breath is largely responsible. The raised temperatures and humidity in the building that houses the soldiers is to blame!

Florida and the Florida Keys

Last year some 10 000 000 visited Florida and 3 000 000 visited the Florida Keys. These two areas are under siege! Is this $91.4 million per day industry worth it?

In Florida

The Everglades National Park – a unique wetland system has undergone massive alterations of the landscape to cope with the tourist influx. Areas have been drained to grow food for the tourists and Florida's aquifer is being sucked dry by the thirsty visitors!

Rubbish and sewage adds to the problems.

In the Florida Keys

Key West is some 13 km long and welcomes 2 million visitors each year!

The Keys as a whole face traffic, transportation, poor water availability and sewage contamination problems. Homes are expensive to buy. And the USA's only living coral reef ecosystem is under stress from visitor pressure.

The impacts of tourism

AQA B	U5
EDEXCEL A	Some U5
EDEXCEL B	U5
OCR A	U4
OCR B	U5

Know both sides of the story!

As much as 90% of revenue leaks from some Caribbean tourist destinations.

The Gambia, Seychelles and Tunisia rely heavily on foreign exchange.

TIM. (Tourist Income Multiplier).

Cost and benefits

Tourism development can be closely linked to a series of environmental and socio-economic impacts as the previous case studies illustrate. Other impacts include:

Economic/social benefits

- Tourists have huge spending power. Tourism stimulates economies and helps the balance of payments.
- Investment in host countries is encouraged. Increased levels of foreign currency circulate in the countries. GDP increases.
- Jobs are created. This job creation has a multiplier effect through the economy.
- Local food and drink is in demand. Farmers benefit.
- Local crafts and goods are also in demand. This is both a money earner and a way of ensuring the craft type is preserved and revitalised.
- As the local infrastructure firms up to cope with the tourists the local standard of living is boosted!
- The host's culture is fortified by interest from the tourists and visitors.
- Healthcare, education and social security develops.
- The role of women is enhanced.
- Horizons and experiences are broadened.

Costs of tourism

- Tourism is linked to reaction to events in a country. Israel has a high reliance on foreign currency from visitors, conflict between Jews and Muslims means the tour operators will not book tours to what will be a UN – Holy Land!
- Development of core and peripheral areas.
- Economic benefits are selective, only a small amount of revenue generated may reach those that need it. Much is returned to foreign investors. Jobs are seasonal and part time.
- Western values and demands are often at odds with those of the locals.
- Tourism may force locals out of their traditional homes or off their ancestral lands, e.g. in Luxor Egypt local 'squatters' are being relocated from Al Gourna so that tourists have an uninterrupted view of the Nile!
- Locals are driven out by rich retirees moving in.
- House, clothes, and food prices are inflated.
- Tourism literally drains countries dry of their water resources.
- Diet changes cause problems, e.g.increased heart disease in Hawaii.
- Honeypot sites suffer chronic congestion problems, e.g. the volcanic Timanfaya National Park, Lanzarote.
- Crime increases.
- Health risks increase, e.g. Kenya and AIDS etc.
- Antagonism – references to dress, behaviour and religion.

CASE STUDY

Tunisia: a case study in mass tourism

Tunisia is the smallest state in north Africa. The World Bank classifies it as a lower middle income country. It became independent from France in 1956.

Tunisia is seen as one of the most stable states in Africa. The ruling party has consistently encouraged private enterprise and a free and open market economy. Overseas investment and foreign tourists are welcomed as its policy is aimed at modernising Tunisia and bringing about economic growth.

Tunisia is one of the most westernised states in north Africa with a tolerant and relaxed attitude towards the lifestyles of foreign visitors and residents.

On independence, Tunisia relied on exports of primary products. Olive oil was the main source of foreign exchange and most of it was sold to France. Tunisia also exported small quantities of phosphates, iron ore, lead and zinc. The government wished to get away from a dependence on cash crops and minerals.

Tunisia is located within the Mediterranean sunbelt. By the late 1950s, large numbers of northern Europeans were descending on the Spanish, French and Italian Mediterranean resorts. Tunisia enjoyed a similar climate to these holiday destinations, with constant summer sun and relatively mild winters. Long, wide beaches were to be found along the east and south east coasts. Tunisia is now firmly established as a Mediterranean sunbelt holiday destination.

Impact

A huge amount has been invested in the construction of new hotels, international airports and many other tourist facilities. The industry directly employs about 2 000 000. By 1997 about 20% of the country's foreign exchange earnings were brought in by tourism. After petroleum and its derivatives, tourism is Tunisia's most profitable export.

A considerable amount of tourist spending remains unrecorded, especially the purchase of souvenirs from peddlers and in the souks (markets). However, the figures do indicate clearly the importance of tourism in the national economy.

Foreign tourists stay mainly in four regions: Tunis and its environs, Hammamet Nabeul on the south side of Cap Bon, the Sousse-Monastir region, and the Isle of Jerba and nearby Zarzis in the south. Each tourist region is served by its own international airport.

Tourism in the coastal areas has further accentuated the seaboard concentration of cities, infrastructure, industry intensive agriculture, better educated inhabitants with a higher employment rate, and political and economic power. 96% of the jobs in tourism are to be found in the coastal areas and this has encouraged migration from the interior. Consequently, economic dualism has emerged in Tunisia with a prosperous coastal region stretching from Tunis in the north to Zarzis in the south and a neglected interior and north coast.

Development

Tunisia is divided into seven tourist planning regions. Considering its late start compared with most of the other destinations, its achievements have been quite impressive.

The industry now has the capacity to accommodate 3 500 000 tourists a year. However, the occupancy rate of the annual hotel capacity has never exceeded 60% and has sometimes fallen as low as 40%. This inability to fill the hotels throughout the year underlines one of the inherent problems of tourism – its seasonality.

Most European visitors to Tunisia are seeking reliable sun and low rainfall, so it is hardly surprising that there is a marked preference for the summer months, with nearly 50% of tourists arriving between June and September.

As long as most of the European visitors to Tunisia are sun-seekers, this uneven distribution of tourists is likely to continue.

Future

National five-year plans reaffirm the government's continuing commitment to tourism. The main aims are to help the balance of payments, diversify the economy and create employment.

CASE STUDY

Environmental impacts and management strategies

Area	*Concern*	*Management approach*
Spain	Coastal, high-rise development	Strict planning controls
Caribbean	Coral destruction, Mangrove destruction	Marine Park designation, coastal development prohibited
	Silting of seabed/algal blooms	Buffer zones replanted along coast land
Himalayas	Depression/rubbish disposal	Many areas run strict permit entry only
French Alps	Damage to delicate flora on piste slopes	Zoning and no new development allowed
Greece	Athens – traffic congestion leads to air pollution	Limit, restrict car usage
	Zante/Loggerhead turtle's breeding ground is being destroyed	Limit access to beach concerned
UK	In the Peak District, ecological damage through footpath erosion.	Reinforce paths and rebuild
	The Hadrians Wall National Trail, a conservation v recreation issue	Issued ticketing, clear tourist management with guide posts etc.

6.4 Sustainable ecotourism

After studying this section you should be able to understand that:

- *sustainable tourism has economic, socio-cultural and environmental dimensions*
- *eco/green tourism is becoming increasingly popular*
- *countries of destination have to accept stewardship of their environments*

LEARNING SUMMARY

Nonsustainable tourism

AQA B	U5
EDEXCEL B	U5
OCR A	U4
OCR B	U5

As tourism grew, the protection of the environment was a low priority. Landscapes were destroyed and congestion and pollution became an increasing problem. Profit became the driving force behind most developments. Nowadays, investment and financing of tourist development projects usually hinges on the sensible and sustainable development of resources; the World Bank, the EU and so on will no longer be linked with environmentally destructive projects. Areas worthy of protection are not just 'natural'. The cultural and built environment is also given due consideration. Consider the article below:

'Take photos but leave no footprints'. The places we visit we irreparably damage!

'An unfair exchange'

No claim is dafter than that tourism helps protect the environment. It is true that it can finance conservation efforts and encourage other countries to preserve the resources they want tourists to see, but few human activities are as destructive as going abroad. Even if we forget the coral reefs smothered in sewage, the savannah woodlands felled for barbeque charcoal and the swamps and streams drained so that we can enjoy showers and flushing toilets, our environmental account would still be firmly in deficit, simply because we have to travel to get there. Air transport is now one of the gravest threats to the global environment, because of the local pollution and disturbance it generates and the vast quantities of carbon dioxide it releases.

Even the oldest argument of all, that travel broadens the mind, collapses when you see what really happens when people go abroad. Tourists, like the customers in old-fashioned shops, are always right. Tour companies try to provide what they expect, rather than showing them what the countries they visit can reasonably provide. For most tourists, the only surprises will be unpleasant ones, when the reality of the countries they visit pricks the bubble in which they travel. Thousands return home even more convinced than they were before that foreigners are dirty, deceitful and dangerous.

Amid the barrage of misleading corporate claims about eco-friendliness, there are one or two companies that have tried to cushion a few of the lesser blows we inflict on the places we visit. Genuinely low-impact tourism has to take place within the range of boats, trains or bicycles. The money we spend must be spread far more evenly: initiatives that encourage tourists to stay in local people's homes, and which invest in local people's projects, offer the best hope of counteracting some of our negative impacts.

Tourism is, by and large, an unethical activity, which allows us to have fun at everyone else's expense. Go, if you have to. But don't pretend you're doing it for anyone other than yourself.

Source: The Guardian, May 1999

Sustainable tourism

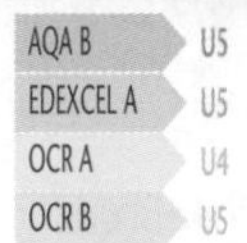

Tourism can be seen as an exploitive industry, depleting the resources at one location before moving on to the next. To avoid this scenario when tourism in a country goes into decline (see figure 6.5, Butler's tourist cycle), efforts must be made to resurrect the situation. But laudable as many efforts are to adapt to sustainable regimes, the huge size, scale, diversity and the swiftness of change in the tourism industry means that 'strategies' are slow to take-off and become adopted (see figure 6.10).

Figure 6.10 *Extremes of tourism*

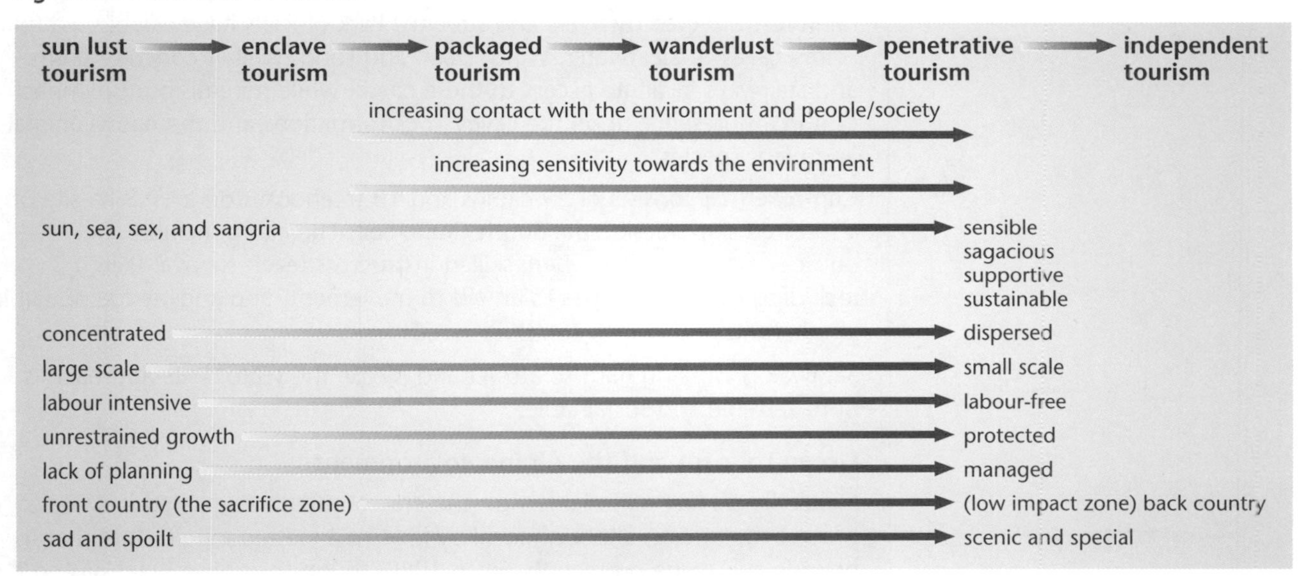

CASE STUDY

The Sarawak experience – ecotourism in action

To build a viable tourism industry around the natural attributes of this wild, remote region, the Sarawak state government in Malaysia has committed substantial resources through its development arm, the Sarawak Economic Development Corporation (SEDC).

SEDC views tourism and leisure as a 'strategic business unit' equal in stature to roads and works, food-based industries, agro-based industries, mineral and mining building materials, human-resource development (i.e. education), and Bumiputra (indigenous peoples), commercial and industrial community-development programmes.

SEDC has invested over 550 million Malaysian ringgits (equivalent to approximately US$154 275 000) to build five international-standard hotels and other tourism properties, plus downtown shopping complexes in Sarawak's capital city, Kuching. The cultural village (a 6.9 ha tourist attraction 35 km from Kuching in the Damai Beach resort district) is a living museum that enables visitors to experience, in half a day, the authentic dwelling styles, arts, crafts, games, foods, music and dance of seven of Sarawak's major resident cultures.

Two of the SEDC resorts, Bukit Saban and Royal Mulu, are located deep within Borneo's interior rainforest and serve as a comfortable base for ecotouristic adventures. Camp Permai, in the Damai Beach resort district, offers a more accessible ecotourism experience in a naturalistic setting.

Bukit Saban is a 50-unit resort built in 1995 on a 7 ha site in the Paku district of Betong, 290 km from Kuching. Accommodation is in buildings that resemble the apartment-style longhouses of the Iban tribe.

In the vicinity of the resort are actual Iban villages where the inhabitants welcome Bukit Saban Resort guests. Bukit Saban Resort also offers guided bicycle, boat, and foot trips in the surrounding terrain.

Royal Mulu Resort was built in 1992 on an 81 ha site along the Melinau River adjoining Gunung Mulu National Park. The entire resort is built on stilts more than 9 m above the ground, well above the high-water mark of the adjacent river's recurrent floods. A net-draped walk-through aviary in the resort grounds harbours a representative assortment of native birds.

Gunung Mulu National Park encompasses 480 km^2 of mountainous rainforest bisected by rivers and streams. Park officials have established four 'show caves' – Clearwater, Wind, Deer, and Lang. Well-lit concrete paths and stairways facilitate access to these caves, while minimising the impact of human presence upon the caves' rock formations and the native animal and plant species.

Camp Permai consists of 29 cabins and 10 treehouses on a 17.8 ha site on a forested slope beside the South China Sea. The camp includes an outdoor-activity centre where skilled instructors teach survival skills, including the use of ropes as an aid to movement, and knowledge of edible jungle fruits and plants.

Sarawak is working hard to attract and please the visitor – its approach is both sensible and sustainable.

Green tourism and the Alpine environment

Switzerland's first national law on the environment was passed in 1902 to protect the forests, although local by-laws had existed long before that to provide avalanche protection. Since 1980, tough laws have been passed on air pollution for industry and motor vehicles, water pollution and waste disposal. Cars and motor cycles have to be fitted with catalytic converters, speed limits are lower than in many other European countries and domestic heating systems using oil have to be monitored for sulphur output.

Switzerland revised its waste disposal laws in favour of the environment, instead of least-cost solutions. Laws include regulations on non-returnable waste, especially in the drinks industry. Such laws are clearly important to the tourist industry. All construction projects have to be assessed to ensure that they conform to all legislation; tourism is specifically mentioned.

CASE STUDY

Ecotourism

AQA B U5
EDEXCEL B U5
OCR A U4
OCR B U5

Well-planned ecotourism conserves fragile ecosystems – whilst at the same time providing income for local companies.

Key principles of ecotourism

1 Maintain the quality of the environmental resource. Develop it in an environmentally sound way.
2 Produce first-hand participation and inspiring experiences.
3 Educate all parties before, during and after the trip.
4 Encourage recognition by everyone involved of the value of the resource.
5 Accept the resource as it exists and accept that this may limit the number of visits involved.
6 Encourage understanding and develop partnerships between governments, non-governmental groups, industry, science and local people.
7 Promote moral and ethical responsibilities and behaviours in the action of everybody involved.
8 Develop long-term benefits to the resource, local community and to industry.

It is possible for tourism to balance its activities with the natural environment.

A topical subject and a favourite with examiners.

Economic advantages of ecotourism:

- a greater variety of economic activities in rural and non-industrial regions
- long-term economic stability
- tourists involved in ecotourism tend to spend more and stay for longer
- demand for local goods and services
- development of infrastructure such as roads, airports and bridges
- increase in foreign-exchange earnings

Benefits of links with conservation organisations:

- donation of portion of tour fees to local groups
- education about the value of the resource
- opportunities to observe and take part in scientific activity
- involvement of local people in providing support services and products
- involvement of local people in explaining cultural activities or relationship with natural resources
- promotion of a tourist and/or operator code of ethics for responsible travel

KEY POINT

Ethical tourism

AQA B U5
EDEXCEL B U5
OCR A U4
OCR B U5

Three questions are central to ethical tourism:

Socially – do we have a right to behave as we wish in our destination resorts?

Economically – should the big operators be able to squeeze rentals so low that owners can barely survive?

Environmentally – do we have the right to intrude?

As tourists we must:
- protect and preserve natural and cultural heritages
- use energy and water resources efficiently
- help to preserve traditions, customs and local regulations
- avoid damaging the environment
- only buy products that demonstrate social, cultural and environmental sensitivity.

The tourist industry must:
- strive for excellence in all aspects
- encourage an appreciation of the environment
- contribute to community identity, pride and quality of life of residents
- protect cultural and aesthetic heritage
- manage waste and pollution appropriately
- support the tourists' quest for greater understanding.

Tourism will remain a growth area for the foreseeable future. Economic development brings more disposable wealth and travel. But what is the future of tourism? (See below.)

Changes in many areas, political, social and economic, will affect tourism in the future. Ensure you keep in touch with all the developments as they crop up. As an A2 geographer you need to be 'up' on your current affairs.

KEY POINT

What future tourism?

A study, commissioned by the World Wild Fund for Nature, found that profitable tourist destinations could be turned into 'holiday horror stories'.

By 2020, visitors to the Costa del Sol could risk contracting malaria as global warming brings more frequent heatwaves, making the country a suitable habitat for malaria-bearing mosquitoes, while increases in summer temperatures to more than 40°C could make parts of Turkey and Greece no-go areas in July and August. The report suggests that operators and countries which rely on holiday-makers for foreign revenue will need to take account of the changing climate when planning new resorts or upgrading facilities.

The tourist industry is not just a potential victim of global warming – it also contributes to the causes of climate change itself.

From the 594 million international travellers in 1996, numbers are expected to rise to 702 million next year, 1.1 billion in 2010, and 1.6 billion in 2020.

Aircraft bring most travellers to their resorts and Air travel is the fastest growing source of greenhouse gas emission, and therefore increases the risk of continued global warming.

The report says:

- a decline in cloud cover over Australia will increase exposure to the sun's harmful rays, increasing the risk of sunburn and skin cancer
- winter tourism will be affected in the Alps and other European skiing destinations from the impact of less snowfall and shorter skiing seasons.
- the southeast coastline of the USA, including parts of Florida, may be threatened by rising sea levels; important wetlands such as the Everglades, could also be at risk
- safari holidays in east and southern Africa may also be affected as droughts and changes in temperature alter the distribution of wildlife
- islands in the Maldives could disappear as they are submerged by rising sea levels.

Sample question and model answer

1

(a) Why do few LEDCs feature in the list of the most popular tourist destinations, and what can those countries do to increase the number of tourists?

Most LEDCs lack the infrastructure found in MEDCs.

Differences in tourism between MEDCs and LEDCs is a commonly examined area.

Plenty of accurate and appropriate detail here.

How to improve LEDC's 'potential'.

The busiest destinations for international tourism are invariably in developed countries, with travellers enticed by a superior infrastructure and a wide range of facilities. MEDCs have a greater ability to cater for large numbers of tourists, with high standards of airports, more hotels and other higher class facilities that one is unlikely to find in an LEDC. Developing countries also lack the fashionable destinations, that belong in the MEDCs, where mass tourism follows strictly to fashion, for example the resorts of Ibiza in the Mediterranean. The tourist industry, if fully exploitable, would offer a lifeline to the developing countries in the southern hemispheres. However, the tourist trade brings with it many problems and hinders the country's chances of exploiting its resources to help build a better infrastructure. One of the largest problems is the undue pressure piled onto the natural ecosystem. Often there is no provision for the local environment, with all available capital being injected into new facilities, especially accommodation. An influx of visitors can cause soil erosion, litter pollution and unbalance the equilibrium of animal populations. There are other social problems brought to a country through the tourist industry. These include the nature of tourist related employment, as it is unskilled, poorly paid and often controlled by Trans National Companies (TNCs) in MEDCs, and leaves workers unemployed out of season and with few rights. Local resources can also be diverted from the native population to the tourist industry, causing hardship among local people. Criminals in tourist areas also direct their attacks at the less secure visitors. Although crimes are often petty, there are serious incidents, many of which are reported, deterring travellers from choosing the LEDC as their holiday destination.

When these problems have been overcome and tourist destinations in developing countries do become popular, much of the profit leaves the country. This is due to the fact that the LEDCs must use international systems organised by TNCs, that are mainly situated in developed countries, to attract visitors. This so-called 'economic colonialism' is said to be evidence of the developed north controlling the developing south.

Tourism can also be very unpredictable, varying with the strength of the economy, safety, cost, alternative opportunities and the stage in family life cycle. The tourist industry in developing countries therefore suffers from many problems, raising the question as to why LEDCs continue to persevere with attracting foreign tourists to their countries. Tourism can bring a number of benefits, including foreign currency, employment, investment and an improved infrastructure. The example of Spain's development of its infrastructure through tourism during the 1960s shows the way that tourism can help a developing country attempt an economic parity with developed countries in a relatively short space of time. In order to allow LEDCs to feature more highly in the world's most desirable holiday destinations, they must ensure that they exploit to the full the resources that they have available to them, many of which the MEDCs cannot compete with. Rich and varied wildlife, climate, beaches, cultural heritage and low living costs are all factors that can be unique to

Sample question and model answer (continued)

developing countries. If the developing countries want to use tourism to help themselves, they must ensure that they use money to improve infrastructure and amenities, preserving their resources, but making the destination more desirable and more suitable in the eyes of the customers from the developed countries.

(b) Assess the economic and social impact in an LEDC that you have studied.

Expect to be asked about economic and social impacts whether your case study is from an LEDC or an MEDC!

Tourism in Kenya is mainly safari and beach oriented. The wildlife-related tourism brought an increased amount of visitors into Kenya through the 1990s, boosting the LEDC's economy. However, many of the areas that are most valuable to the tourist trade are the wildlife-filled parks, that have been inhabited by peoples like the Masai for hundreds of years. The Masai people have coexisted with the natural ecosystem as a pastoralist group.

The introduction of the tourist industry into Kenya has led to the Masai being forced to change their culture, to update their values, aspirations and beliefs, as they become steadily influenced by a modern cash economy. The Masai are increasingly involved in an economy where they need to earn money to buy goods and services, that previously they would have provided for themselves. Capital for food, medical bills, school fees and consumer goods must be raised and many Masai have become involved in tourism, opening bars and shops and providing accommodation and transport. Others lease their lands to building contractors. The Masai have been able to take advantage of the tourist trade in these ways, and create a profit. Many of the Masai are losing their traditional values and developing ideas of individualism. The Kenyan government is pleased by the surfacing of these entrepreneurial skills, stating that the Masai are modernising and contributing to the national economy. Others are less pleased, seeing the Masai losing their land, their rights, their culture and their livelihood.

Clear focus on advantages and disadvantages of tourism.

For many years the Kenyan tourist trade ignored the indigenous population, who became more disadvantaged as their lands were exploited. As a result of tourism consumer prices rose and vegetation was destroyed with the construction of new roads. Population also increased in tourist areas, creating adverse environmental and social effects, including the breakdown of communities. During the 1990s an increasing number of Masai benefited from tourism. The infrastructure of Kenya was developed through business and political links, but this threatened local cohesion, although the small minority were becoming very wealthy. Those Masai tribes continuing with traditional pastoral methods were also used as bait for tourists. Many groups managed to combine livestock rearing with ecotourism, with others involved in wildlife projects and community development projects. The Masai social structure is also breaking down. A land-owning elite has formed, disinterested in the Masai people but interested in capitalist Kenya and desiring a greater share of Kenya's riches. A lack of formal education and entrepreneurial skills from the Masai communities means that they are unable to claim what legally belongs to them. The traditional land uses in Kenya are therefore being integrated with tourism.

Sample question and model answer (continued)

Kenya must retain its natural ecosystems if it is to continue to attract visitors, and the tourist industry therefore must work with the remaining traditional Masai groups to organise a sustainable management of the wildlife. In turn, the tourist industry must reward the groups, otherwise they will simply revert to intensive, commercial farming methods, that will spoil the ecosystem and Kenya's chances of development through tourism.

Comment

Super in every respect: fantastic use of case study in part (b), its effectiveness is maintained through the whole answer. Probably based on first-hand experience. The essay 'flows' and uses the language of the subject. A high level response.

Practice examination questions

Resource A:

Growth in numbers of foreign tourists and income from foreign tourists, World, 1950–1995.

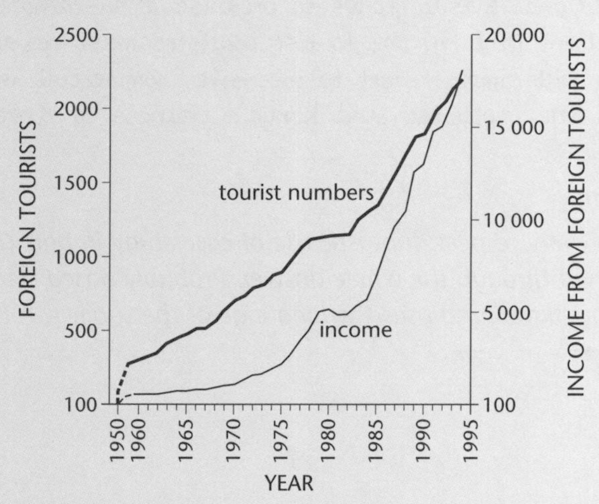

Source: Worldwide Fund for Nature

Resource B:

Taming the Wild Creatures

Free ranging animal populations damage Kenya's human habitats

To many of the tourists who visit Kenya each year, Peter Oyuka would seem to be a lucky man. Living in the shadow of Mount Kilimanjaro, near the famous Amboseli National Park, he sees each day the kind of wildlife that foreigners pay thousands of dollars to glimpse just briefly. Yet Oyuka, a migrant farmer and father of ten, does not feel fortunate. His small plot of 0.6 ha is regularly destroyed by zebras and elephants straying from their traditional grazing grounds. Last year a neighbour trying to protect his meagre crop of maize was paralysed from the waist down after being seized and thrown by an angry elephant. 'They are dangerous to us and our livelihood', says Pyuka gazing at a field of flattened maize. 'If I had the power, there would be no wildlife here.'

Ensuring that tourist dollars trickle down to Kenya's poor people remains a challenge. Those people most troubled by wild animals see little of the profit they generate. Though wildlife tourism is Kenya's biggest source of foreign exchange, profits are usually taken by the central government or invested into national parks not used by most Kenyans. The Masai Mara park has long been controlled by a county council that routinely diverts an average of 40% of gate receipts into its own coffers. Nor does setting up private reserves to keep the animals from farmland guarantee that those most harmed by the animals will benefit. The director of one private reserve worries that if his venture starts to make a profit from tourism, central government officials will step in and take over the operation. The Director of the Kenya Wildlife Service has staked his reputation on the promise that such seizures will not take place.

1

(a) Study Resource A which shows changes in international tourism from 1950–1995.

(i) Identify when and how international tourism was affected by a period of severe recession. [2]

(ii) Describe how and why the international tourist industry changed between 1950 and 1995. [5]

(b) Study Resource B which details some of the consequences of tourism in Kenya. Critically evaluate the Kenyan authorities' policy of encouraging wildlife tourism. [9]

(c) With reference to a case study of tourism:

(i) Describe the tourism management policy adopted. [4]

(ii) Evaluate the success of the policy in relation to the creation of revenue and employment, and to minimising the negative impacts of the industry. [15]

CCEA Specimen

Chapter 7

Agriculture and food supply

The following topics are covered in this chapter:

- *What is agriculture?*
- *Issues in agriculture*
- *Organised industrial agriculture (OIA)*
- *Future trends*

7.1 What is agriculture?

LEARNING SUMMARY

After studying this section you should be able to understand that:

- *agriculture is a system that farmers can manipulate to enable them to maximise output and profits*
- *agriculture operates at a variety of scales and levels and can be classified in a number of ways*
- *a variety of factors influences the decisions made by farmers*

Defining agriculture

AQA B	Some U4
EDEXCEL A	Some U4
EDEXCEL B	U5
OCR A	U5
OCR B	U5
WJEC	U5

Know your definitions.

Nearly half the world's population is involved in agriculture, 96% in LEDCs.

Agriculture is the biggest user of land.

This sort of summary could feature as an essay plan!

There is no precise definition but agriculture is generally taken to be 'the use of animals and plants for the production of food and other materials for human consumption'. Over the last century the face of agriculture has changed many times: machinery, HYVs (High Yielding Varieties – of seed), genetic modification, fertilisers, agro-chemicals, pesticides and herbicides have all influenced and contributed to food production. The situation is now turning full circle upon itself, with the organic food revolution. All of this is happening against a backdrop of changing and variable government and multinational involvement.

Agriculture can be studied from a number of perspectives:

- **issues** – e.g. set-aside, hedgerow destruction, eutrophication
- **processes** – e.g. food mountains, CAP, environmental damage
- **changes** – e.g. intensity of cultivation, changing tastes
- **interactions** – e.g. physical and economic interactions
- **patterns** – e.g. of different crops.

The variable purposes of agriculture

Producing food is just one of the purposes of agriculture. There are a range of others and these are shown in figure 7.1.

Figure 7.1
The purposes of agriculture

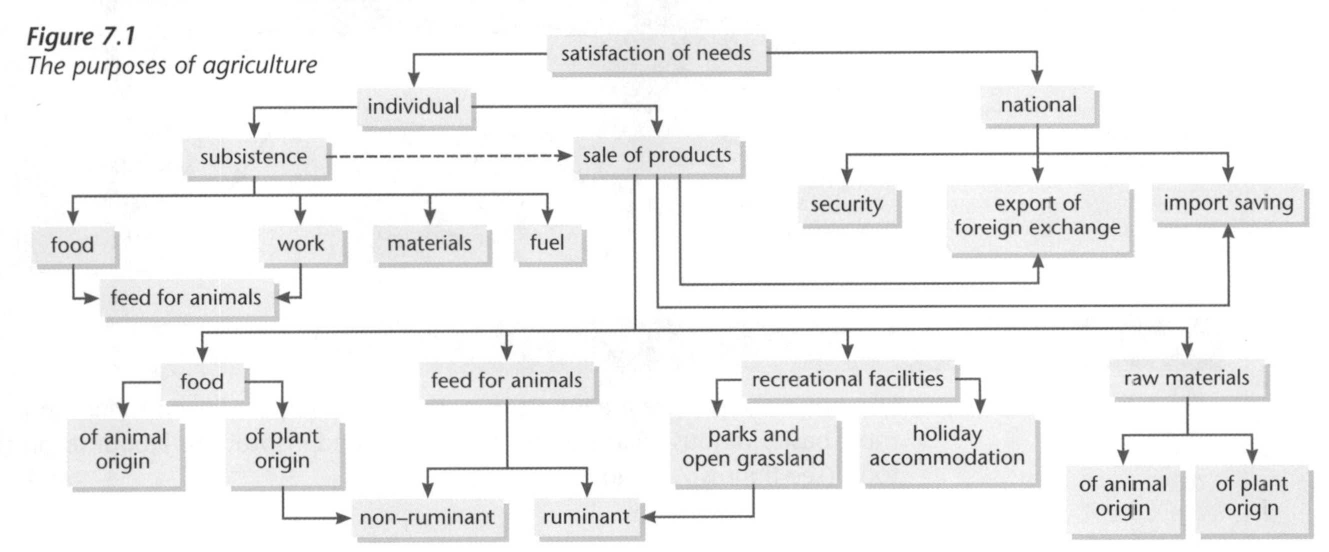

The difference between agriculture and industry

Agriculture	Industry
• Uses large amounts of land	• Tends to occupy small sites
• Strongly influenced by the physical environment	• Generally has little effect on production.
• Environmental impacts are huge	• Environmental impact is surprisingly less than in agriculture
• There is a time lag between planting/ investment and yield	• Raw material comes into the factory, is made into something and leaves. There is no time lag in production
• Less speculation, in general	• Highly specialised
• Small production units are common	• A mixture of different sized units
• Less corporate interference	• The corporate nature is strong
• Location is fixed	• Variable footloose location
• Most farmers (LEDCs) consume the food that they produce	• Trade and markets are important

Frequently in essays you have to demonstrate differences between systems.

Importance of agriculture

- It employs millions around the world. In fact about 1 : 2 are involved in agriculture! In terms of workforce, the biggest are in the LEDCs, in some areas 90% are involved in agriculture. In MEDCs as few as a half of 1% can be involved in agriculture!
- It makes a strong contribution to economic output, wealth/GNP/GDP.
- Being so important ensures the issues it raises are always in the news.

Classifying agriculture

Traditionally farms were classified by the dominant enterprise: be it dairying, arable production and so on. It is now recognised that a whole gamut of characteristic prefixes and suffixes can be used and applied to farming systems. Some common classification criteria are shown below.

Traditionally agriculture has focused on this aspect. However, most new specifications include this as a minor element in their agricultural 'orders'.

It would be wise to learn a case study for each classification.

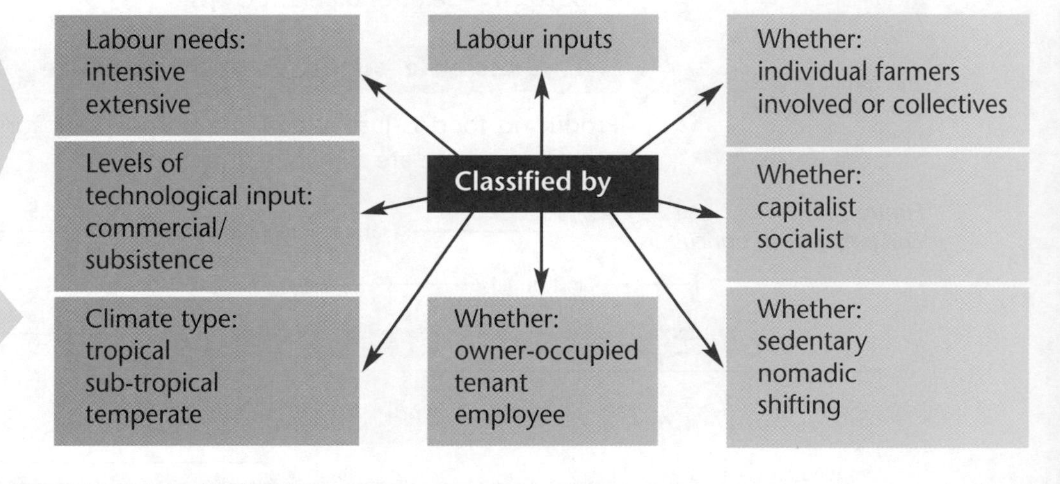

The agricultural system

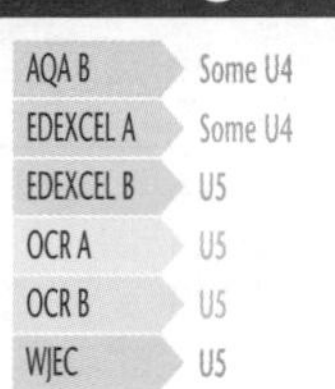

Like any system there are inputs, processes and outputs linked to farming. At its most basic, various chains of people, groups and institutions labour to produce food (see figures 7.2 and 7.3).

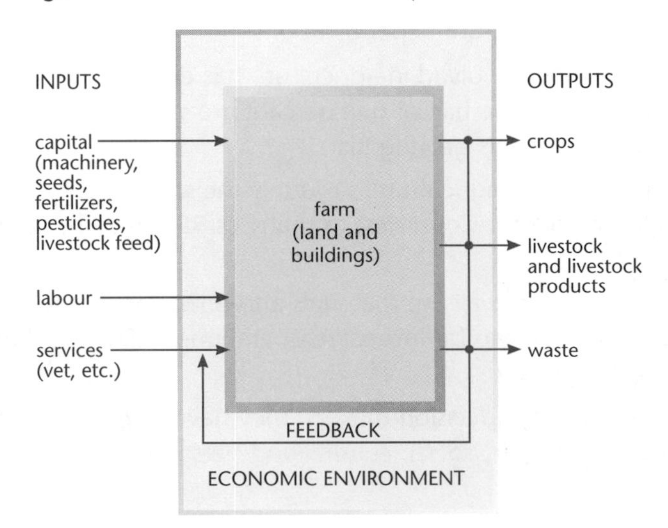

Figure 7.2 A farm as an economic system

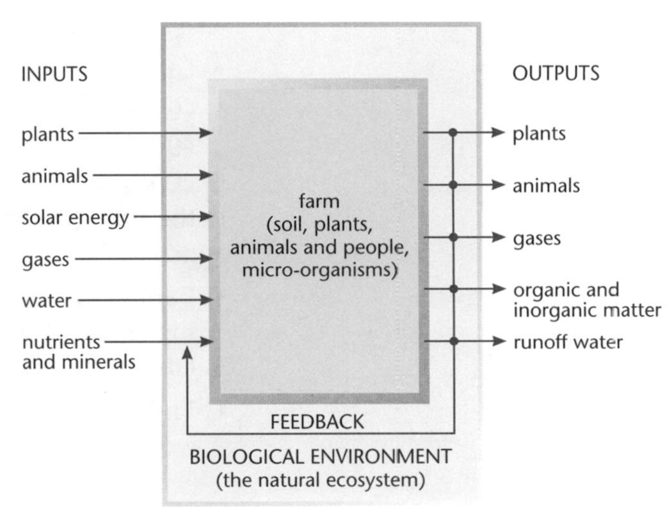

Figure 7.3 A farm as an agroecosystem

Worth remembering for the exam.

If agrisystems are balanced in terms of inputs and outputs, such systems are said to be sustainable. Organic farming attempts to do this.

Influences, impacts and factors involved in agricultural production

Influences

- **Climate:** Precipitation is probably the most important climatic characteristic.

The significant factors are:

- annual amounts
- seasonal distributions
- variability.

Irrigation can overcome many of these problems.

As **temperature** varies with latitude, influences tend to be at global, regional and local scales.

- **Soils**: these vary in their physical and chemical characteristics and this influences agriculture.

All of these factors, influences and impacts would make a super essay.

Physical factors	Chemical factors
• soil moisture budgets (relates to water holding capacities)	• soil acidity – the best soils are between pH 6.5 and 7.5
• soil texture (relates to drainage)	• soil nutrients – the most productive soils have a high nutrient status
• soil structure (relates to the way humus and minerals stick together)	

Impacts

- **Environment** Huge areas of the globe are farmed: this has a distinct and profound effect on the environment.

Positive	Negative
• development of the agricultural landscapes of fields, hedgerows and trees	• land degradation (soil erosion, salinisation and desertification)
	• reduction in biodiversity
	• deforestation
	• water pollution

Key points from AS

- **Soils and ecosystems**
Revise AS pages 78–89
- **The dynamics of population**
Revise AS pages 110–124

Factors

Like any economic activity agriculture depends upon:

- **Land** In MEDCs the number involved in agriculture has declined; in some counties in the UK less than one half of one percent are still involved in agriculture, and the workforce is greying fast.

In LEDCs the number involved in agriculture is roughly the same county to county; in the HIPC (Highly Indebted Poor Countries) as many as 95% may be involved in agriculture.

- **Capital** In farming this includes all the materials and financing needed to run the farm. MEDCs tend to be **capital intensive** operations. LEDCs tend to be **labour intensive** operations.
- **The farmers** Most farmers are decision-makers, they have to decide what will work to achieve their economic, social, cultural and personal goals.

7.2 Issues in agriculture

LEARNING SUMMARY

After studying this section you should be able to understand that:

- *commercial farming is very productive, but that this comes at a cost*
- *population increases drive increased agricultural output*
- *governments can affect output*

The current situation

AQA B	Some U4
EDEXCEL A	Some U4
EDEXCEL B	U5
OCR A	U5
OCR B	U5
WJEC	U5

Issues are the main thrust of agricultural geography in the new specifications.

Agriculture has undergone many changes and revolutions over the last century or so, to ensure that food production has kept pace with the world's booming population, e.g. HYVs, fertiliser and pesticide usage, bio-technology in the form of GM (genetically modified) foods and so on. But at the same time this has been going on there has also been an increase in the numbers of people that want 'chemical free' or **organic foods**. The other major change has been the concentration of power into the hands of small numbers of multi-national firms, concerned with the growing, marketing, processing and packaging of foodstuffs.

Population increase and food supply

There are at present some 800 million for whom governments are unable to provide an effective food supply.

It is population growth that has caused many problems for agriculture.

Over the last 50 years population has more than doubled. Meeting the demand created by extra mouths is the responsibility of individual governments. Governments and other groups have dealt with this issue in a variety of ways.

- **By importing food** This has inbuilt problems, especially when the producer country has years of shortage itself and of course during years of conflict.
- **Government policy** After the last war opinion was that the UK would not be able to support its growing population and it also recognised the vulnerability of its supply chain. Wide scale industrial farming was adopted, this drove production up, with some considerable environmental cost. However, the UK farming system is now extremely efficient.
- The maintenance of food supply was also behind the **adoption of a common agricultural policy** by Germany and France in the 1950s. Its influence continues to affect the members of the EU ensuring that the smaller farmer's production and prices he gains for products are maintained.

Set-a-side occurred as way of curbing spending initially. Now it is sold as being sustainably sound!

Plenty in the press about CAP, always!

Larger intensive producers, have gained most on the scheme!

Reform of CAP is called Agenda 2000. Present EU agricultural budget is about $43 billion.

KEY POINT

Agricultural policies and subsidies

Food security is the driving force behind subsidised farming; payment based on production, low-cost loans and cheap farm chemicals (fertilisers/ pesticides) are the typical incentives used by governments. Some $350 billion is subsidised by MEDCs/year. Subsidisation has promoted intensive methods and it has carried on despite 'free market' pressures to rid the agricultural world of them.

European Common Agricultural Policy

In the EU it is recognised that subsidies are costing Europe too much. The greatest challenge to the EU is the reform of the Common Agricultural Policy (CAP). Those that contribute to GATT have insisted that subsidies and tariffs are firstly reduced and then removed to encourage trade. These GATT rules are meant to help poor countries, but it is expected that the main groups to gain will be the multi-nationals.

How does CAP work?

Up to the early 1990s most producer support under the Common Agricultural Policy came from measures designed to maintain producer prices for agricultural commodities at above world market prices. This was achieved by:

- a quantity of the product being removed from the market and placed into intervention stores when prices were low, to effectively stop prices falling below certain levels
- placing levies on imports from countries outside the EU to prevent them undercutting domestic prices
- making export refunds available to enable EU agricultural commodities to be exported at world market prices.

Reform of the CAP

During the 1980s the CAP became increasingly costly as production levels increased, it therefore became necessary to reform the CAP. The reformed CAP came into existence in 1992. There was to be a move away from production-related subsidies, which were needed less as the EU had become self-sufficient in many commodities. There was to be a move towards direct subsidies subject to quotas. There was also to be an encouragement of more extensive production by increasing subsidy payments if stocking rates were below certain levels.

World trading agreement

In 1995, a new world trading agreement on meat and other agricultural products was reached. Under the World Trading Organization (WTO) (formally known as GATT) agreement, the EU was committed by the year 2000 to substantial reductions in exports made with the help of export refunds. Further reforms of the CAP taking place in 2000 and beyond; these continue the reform process begun in 1992.

Some problems with CAP

- CAP displaces instability from the EU into the wider, world market
- CAP also conflicts with Article 110 of the Treaty of Rome, which states that member states should aim to contribute to the 'harmonious development of world trade'
- it may also threaten the USA's agriculture and lead to a trade war
- it also generates resentment among Eastern European countries and LEDCs.

The furore looks uncertain to go away as the USA and Cairns group have both agreed that the reforms don't go far enough. They argue that the EU must open their markets to agricultural exports from the world's poorest countries too, since agriculture is the main source of income, and closed markets put them at an unfair competitive disadvantage.

- Governments **have actively encouraged changes in diet.** These changes must be reflected in governments' approaches to food production.

Increasing consumption of animal protein is leading to many problems in LEDCs.

By midway through this coming century agricultural production is expected to have tripled or quadrupled to keep up with demand from the world's burgeoning population. Both MEDCs and LEDCs will have to maintain food supply against growing future demand without substantially further damaging the environment or people's health.

GM foods could be feeding the biggest percentage of people by the 21st century.

KEY POINT

FAO forecasts sufficient grain production for 2000 – but what of the future?

The UN Food and Agriculture Organization (FAO) predicted global grain production of 1.895 billion tonnes this year, only slightly lower than last year's record 1.908 billion tonne yield. According to the UN agency's initial projections, this year's yield will be sufficient to meet consumption needs in the 2000–2001 period. But grain stocks will remain below the minimal level needed to guarantee global food security.

FAO stressed that its estimate was only provisional, because many of this year's harvests have not yet been planted while others are in the initial stages of growth.

FAO says the possibility of continuing crop losses in the future cannot be ruled out, especially in a number of southern hemisphere countries affected by unpredictable climatic conditions related to the El Niño weather phenomenon. El Niño's impact on agriculture could revert the positive trend in global grain stocks. This year 37 countries are facing food emergencies, compared to 31 in late 1997, due to the effects of El Niño in Asia and Central America.

But the worst food shortages will be seen in Africa, the consequence of adverse climatic conditions and civil unrest.

Source: The Guardian Online 2000

7.3 Organised industrial agriculture (OIA)

LEARNING SUMMARY

After studying this section you should be able to understand:

- *the reasons for and the effect of the rise in OIA approaches on both arable and livestock production*
- *that the use of toxic chemicals to fertilise and to ward off pests can be harmful to the environment*
- *that biotechnology has an increasing effect on crop and livestock production*
- *that OIA can affect both the environment and our health*

The global food system

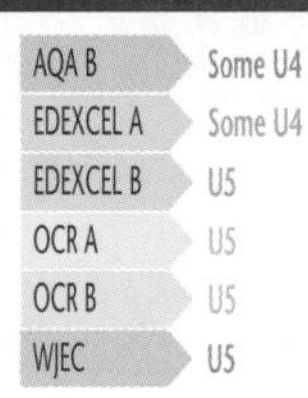

'What has emerged today is a food system for those who can afford to buy, which is largely the people living in the MEDCs – that is around a quarter of the world's population. It uses an industrial approach to agriculture and food production, is highly productive in response to high inputs and overcomes seasonality for all foods. It draws on produce from around the world and by using a mixture of trading and preservation techniques, enables a wide range of foodstuffs always to be available.... Commodities are produced, traded and transformed, bought and sold, in a market whose reach has extended from a largely local level to an increasingly global stage. It is the market in which the actors seek to control their costs, their production or their marketing practices, as closely as they can. They want to minimize their uncertainties and costs and maximize their returns...'.

Source: Food Systems, *by Tansey and Worsley*

Features of OIA

AQA B	Some U4
EDEXCEL A	Some U4
EDEXCEL B	U5
OCR A	U5
OCR B	U5
WJEC	U5

Most money is made in the processing portion of agriculture.

There is little doubt OIA could support the biggest proportion of the world's population.

Specialisation

OIA has brought an end to peasant agriculture in the MEDCs. In the UK, for instance, less than 1% of the working population over the country are maintained by agricultural activity (with a further 7% involved in food processing activities).

It is always useful to plan and write out points like these when you revise.

CASE STUDY

Issues in agricultural employment in the EU

- **A fall in agricultural employment and north-south differences**
  On average, agriculture accounted for 5% of EU jobs in 1997, rising to over 10% in Greece, Portugal and Ireland. In countries where the more labour-intensive Mediterranean-type production predominates (Italy, Spain, Greece and Portugal), farming accounts for 9% of jobs on average. Arable farming and animal husbandry, which are more common in the countries of northern Europe, require less labour (agriculture accounts for an average of 3% of employment in Denmark, the Netherlands, Finland, Spain and the UK). There are differences between northern and southern Europe in terms of the age distribution of heads of holdings. The proportion of elderly farmers is generally higher in the Mediterranean countries, with nearly one in two farmers over 55 years of age as opposed to only 1% in Germany.
- **In relative terms, employment in the agri-food sector is highest in Denmark and Ireland**
- **Structural weakness in the agricultural sector**
  Analysis of the levels of agricultural and agri-food employment highlights the special situation in Ireland, Greece and Portugal, where agriculture plays a dominant role in terms of processing and marketing jobs.
- **Two agricultural jobs and one agri-food job per hundred inhabitants**
  The number of farmers per hundred inhabitants has tended to fall while the number of agri-food jobs per hundred inhabitants has remained static.
- **Family labour is the keystone of agriculture**
  Family labour predominates in agriculture, accounting for four out of every five jobs in 1995. The number of non-family workers is highest in the United Kingdom and Denmark. The fall in the number of farms and their increase in size has not necessarily led to a corresponding increase in the number of paid workers. Farmers are becoming more educated, in the UK 11% have undergone Higher Education.
- **Part-time work predominates in southern Europe**
  Part-time work is much more widespread in the countries of southern than northern Europe. The large number of seasonal activities in southern Europe is one explanation for this high level of part-time work. There would appear to be underemployment, perhaps due to structural weaknesses in the agricultural sectors in these countries.
- **Agriculture remains a driving force for economic and social cohesion**
  Overall, agricultural employment is clearly falling in Europe. Agriculture is doomed to play a secondary role in the process of economic and social cohesion in certain regions, and particularly rural regions. Even as a minority in the countryside, farmers are still the main managers of the land and agricultural work largely determines

> the degree of attractiveness of these regions, particularly where the landscape is concerned.
>
> - **Environmentally friendly farming methods generate employment** In Great Britain agri-environmental measures are leading to a small increase in agricultural work and a substantial increase in work by firms.
>
> CASE STUDY

OIA has also brought an end in MEDCs to the self-sufficient community. Nothing was wasted when traditional farming methods were used.

Tea, coffee, cotton.

Colonial monoculture is especially vulnerable to pests and diseases.

It has also meant, and enforced for many countries, a rise in **monoculture**. 'Colonial' monoculture was imposed by many of Europe's colonisers on LEDCs well before the 1900s. On the whole these plantation crops have very little effect on the host GDP and provide no 'food' support.

In some LEDCs monoculture has been put in place by the countries themselves to ensure that their population can be fed; but also to ensure that yields can be sustained year on year.

Despite year on year use, soil still sustains yields with no special additives.

> **GM rice – the rice that could feed the world**
>
> Genetically modified rice could boost yields by up to 35%, solving the world's impending food shortage. Rice stocks are dwindling and there are fears of famine as the world's population increases. Half the population, including almost all of east and south-east Asia, is dependent on rice.
>
> Genetic material from maize can be inserted into rice, boosting the rate of photosynthesis so the plant is able to produce more sugar and increase grain yields. Farmers must consistently produce an extra 6.7 million tonnes of rice a year, using less land and less water, just to maintain current nutrition levels. In India and Bangladesh the idea of having to buy seed each year from giant companies has brought serious resistance to GM technology, and GM companies have been heavily criticised for not helping the developing world more. Supporters of genetically modified crops pointed to the potential benefits of another GM rice announced in January, called golden rice because of the colour caused by the modification. This adds vitamin A to the grain and could cure the vitamin A deficiency of 124 million children worldwide, a deficiency causing blindness.
>
> All the crops commercially developed so far, such as soya, maize and oil, seed rape, have been aimed at rich farmers in the USA, Europe and Brazil, adding to surplus stocks rather than feeding the world's hungry.
>
> KEY POINT

Mechanisation

Changes brought on by two world wars meant the use and adoption of machinery and new technology was inevitable to ensure food supply, and have ensured that the USA and Canada have become effectively the 'bread basket' of the world.

The Green Revolution has mirrored the productivity of the intensified MEDCs' agriculture. Gains have been made not by mass mechanisation, but through the use of fertiliser, pesticides, and above all else the use of HYVs of seed.

Intensification

The intensiveness of production has ensured a quadrupling of production in MEDCs.

On the huge open prairies of North America this has led to massive consolidation of farms into larger and larger units, using less labour. The huge levels of mechanisation needed to run and manage these huge 'estate' farms means that increasingly financial groups and institutions have a considerable stake in, and are a driving force for farmers, ensuring continued high production and profitability.

Most farmers in MEDCs have a heavy debt burden, brought about by the need for massive machinery to farm their vast estates. Profit, necessarily, has to be at the forefront of their operation.

Commercialisation

A further consequence of OIA has been the rise of the controlling large company – the rise of **agri-business**. These companies not only influence what is grown, but also what seeds and fertilizer are used! Undoubtedly the industrial approach is better suited to the more developed world, with their big fields, equitable climates and big machines. But even in LEDCs the influence of industrialised agriculture is increasing. This against a background of steep terrain, sometime arid and sometime tropical climates, environments that on the whole are unsuitable for such techniques.

Organised industrial livestock farming

The industrialisation of agriculture has had particularly dramatic effects on the farming of livestock; and enormous impacts on meat supplies.

Mass production methods

Plenty of press coverage available for this aspect.

600 million broiler chickens are reared/yr in the UK. At slaughter they are 6 to 7 weeks old.

Many farmers now consider the rearing of animals, for slaughter, in individual small units too inefficient. Many treat animals as units of production (ignoring basic needs such as exercise, fresh air and wholesome food) and have designed systems for turning huge numbers of animals into vast quantities of cheap meat extremely quickly. Battery units and broiler sheds for chickens, sow stalls and farrowing crates and poultry sheds. Being crated, cramped and confined is the price that many animals pay to ensure that our plates are filled.

Abuse of drugs

2000 human deaths/yr are related back to antibiotic misuse. As a result more virulent bacteria harmful to health appear.

New technology is further extending livestock productivity: injections of bovine somatotrophin can increase milk production in cows, clenobuterol is used to boost the rates of growth gain in cattle for beef. With factory farms attempting to counter the effects of intensive confinement and its associated disease-ridden conditions by administering high doses of antibiotics and other drugs to animals. The most commonly used are penicillin and tetracycline. The squandering of these important drugs in livestock production is wreaking havoc for physicians in the treatment of human illness.

As well as suffering from anaemia, influenza, intestinal diseases, mastitis, metritis, orthostasis, pneumonia and scours the animal's behaviour is also affected.

CASE STUDY

Battery hens!

300 million egg-laying hens in the USA (30 million in the UK) are confined in battery cages – small wire cages stacked in tiers and lined up in rows in huge warehouses. The USDA recommends giving each hen 4 inches of 'feeder space', which means the agency would advise packing 4 hens in a cage just 16 inches wide. The birds cannot stretch their wings or legs, and they cannot fulfil normal behavioural patterns or social needs, they suffer from severe feather loss, and their bodies are covered with bruises and abrasions. All laying hens have part of their beaks cut off in order to reduce injuries resulting from excessive pecking, an aberrant behaviour which occurs when the confined hens are bored and frustrated. Debeaking is a painful procedure.

Laying more than 250 eggs per year each, laying hens' bodies are severely taxed. They suffer from 'fatty liver syndrome' when their liver cells, which work overtime to produce the fat and protein for egg yolks, accumulate extra fat. They also suffer from what the industry calls 'cage layer fatigue', and many die of egg binding when their bodies are too weak to pass another egg.

A hen will use a quantity of calcium for yearly egg production that is greater than her entire skeleton by 30-fold or more. Inadequate calcium contributes to broken bones, paralysis, and death in battery hens. After one year in egg production, the birds are classified as 'spent hens'. They end up in soups, pot pies, or similar low grade chicken meat products where their bodies can be shredded to hide the bruises from consumers.

For every egg-laying hen confined in a battery cage, there is a male chick who was killed at the hatchery; male chicks of egg-laying breeds are of no economic value.

CASE STUDY

In Norfolk 'chicken pooh' power stations have been built to deal with the waste produced!

The upshot of growing awareness of the conditions in which livestock are kept means that many thousands are turning to vegetarianism as a matter of conscience. Other concerns to do with hens are the massive amounts of waste they produce and its effect on the climate and global warming.

The cost of OIA

EDEXCEL A	Some U4
EDEXCEL B	U5
OCR A	U5
OCR B	U5
WJEC	U5

The march of OIA has been just about unstoppable. Over the last 40 to 50 years, grain production has doubled, even trebled in some countries and resulted in a range of environmental and health problems.

Below are offered some of the problems and changes that result from the increasing grain harvest:

Nitrogen phosphorus and potassium are very necessary!

113 million tonnes of fertiliser is applied annually to fields around the world.

- **Fertilisers** These effectively extend the life cycle of the soil by artificially adding nitrogen. The application of fertiliser is often overdone, the excess being washed into watercourses causing **eutrophication** and contaminating water more generally with nitrates. Nitrates are linked with a range of disorders including '**blue baby syndrome**'.
- **Seeds and plants** There was a time when all countries and areas grew their own crops, ones suitable for the areas and region in which they lived. However, in today's '**industrial economy**' large **economies of scale** and production could not be achieved using such a system. Today, less than half a dozen high yielding, pest- and disease-resistant varieties are grown; and they supply nearly three-quarters of the world's calorie intake. To enable these varieties to 'provide' as they do, considerable '**genetic engineering**' has been undertaken. Some consider that this **genetic modification (GM)** is intrinsically wrong or 'unnatural'. Others are uneasy about what GM means for our relationship with the natural world. For many it is a step too far, concerned as they are with the consequences, rather than the rightness or wrongness of manipulating genes.

Many countries have signed the Convention on Biodiversity – which enshrines the precautionary principles in relation to transboundary movements of GM crops.

Food security in HIPCs and highly populated countries

There is a world food crisis. Currently at least 800 million people are undernourished. One-third of the world's children go to bed hungry. In an ideal world resources would be redistributed, poverty reduced and food security ensured. Technological 'fixes' can help, and though the GM route can't solve the problems permanently it is being utilised.

Asia leads the race to develop GM crops

India and China, the world's two most populated countries, are leading the Asian race to develop genetically modified crops to feed their growing populations.

Despite serious concerns in Europe over the long-term environmental and health effects of genetically modified food, field trials on GM crops are going on in Asia, with biotechnology research in both India and China on the rise because of government support. Leaders in Asia are increasing research on biotechnology. They are moving toward trials on a variety of products that will benefit their countries not only on maize but rice,

CASE STUDY

GM is topical – ensure you know all there is to know about it!

papaya, cassava. China has been working with biotechnology for about 12 years and genetically modified cotton has been planted on 150 000 hectares of land in China. China is at present spending about $80 million a year for countrywide research and development on GM crops.

India's Department of Biotechnology urged the government last year to provide $3.44 billion for a 10-year plan to develop biosciences.

CASE STUDY

Super summary here – know and understand it!

KEY POINT

The potential opportunities and risks of GM crops in developing countries

Socio-economic

- Higher agricultural yields and labour productivity
- Increased farm income, given appropriate market conditions
- Reduced costs for producers, e.g. from reduced dependency on external inputs such as fertilisers and pesticides, and increased effectiveness of herbicides
- Cheaper products for consumers, e.g. jeans made from self-dying blue cotton
- Enhanced assets of poor people, where they have security of tenure, e.g. by enabling cultivation on land that was previously considered useless for farming, such as saline soils
- Cheaper staple foods for net food-importing developing countries

- Corporate control is increased over seed and agro-chemical markets (through patenting, monopoly production), at the expense of poor farmers
- GM varieties extend the range of crops that can be produced in northern temperate zones and these substitute for developing-country export products
- GM crops are introduced faster than the development of the regulatory capacity in developing countries
- Potential opportunities by-pass smaller farmers, due to lack of investment in research on relevant crops/applications
- Biotech companies bear little liability for any damage to the environment or public health resulting from use of the technology
- Large-scale GM crop farming makes GM-free and organic farming *de facto* impossible
- GM crops and related technology packages considerably reduce production costs but are only adopted by larger farmers. Which increases problems of uncompetitiveness for small-scale farmers

Environmental

- Reduced pressure on the environment, principally from reducing use of pesticides and herbicides
- GM plants can be developed for removing toxic chemicals from soils
- Production of biodegradable plastics

- Further losses of biodiversity from monocultures
- Alien genes (including 'terminator genes') transfer from GM crops to other varieties of the same crop and to other species, with unknown effects
- Increased resistance of weeds and pests to agrochemicals, resulting in increased use
- Decreased natural soil fertility (through reducing the activity of nitrogen-fixing bacteria)

Health	
• Elimination of allergens and toxic substances in crops • Production of vaccines	• Increased allergens, antibiotics rendered ineffective and viruses spread across species (e.g. from plants to human gut bacteria)
Consumer choice	
• Improved quality and shelf-life of fruit and vegetables • Improved flavour, texture, and nutritional content of food crops • Lower prices for products	• Threat to choice to make informed purchasing decisions based on social, ethical, religious, dietary, and environmental preferences

Source: Oxfam

KEY POINT

In LEDCs it has led to salination and water logging.

- **Water** Much of the water used by man finds its way onto the world's crops: irrigation is carried out on 25% of the world's cropland. Agriculture must in the future balance its needs against the needs of growing populations.

1 tonne of fertiliser = 1600 ltrs of fuel to make it! 10 units of fuel = 1 unit of food energy!

- **Fuel consumption** Farmers are big users of fossil fuels, especially in MEDCs where high mechanisation uses masses of agricultural diesel (Red diesel as it's called).
- **Pesticides, herbicides, fungicides and insecticides (biocides)** – one rationale put forward is:

20 to 30% of harvested food is spoilt by pests and disease. A bigger proportion never makes it to harvest!

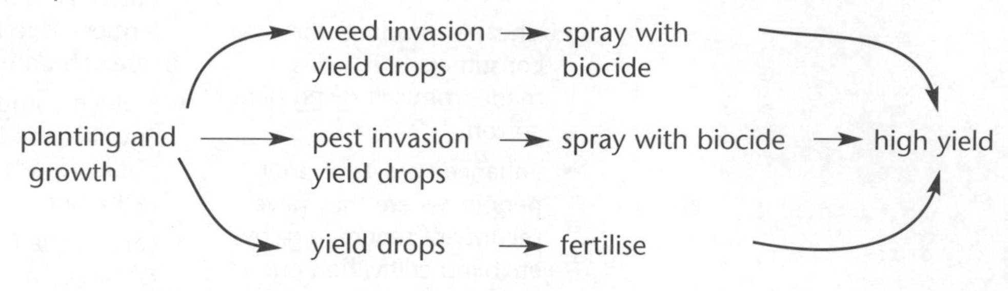

Biocides

The world market for pesticides is worth about $35 billion.

Problems with biocides are mostly confined to LEDC usage:

- DDT (an organochloride) is still used, as are organochlorines that will persist in the soil. The link between accumulations in food chains and deformities, deaths and illness is well documented.
- Poor education ensures that biocides are not being properly applied or effectively used in many LEDCs.
- Purchase costs are high. In many countries and instances better 'fixes' are invariably more sustainable and of greater long-term benefit to LEDCs.
- Chemicals banned in the West are often found as toxin residues in/on imported LEDC products, e.g. bananas.

Within most agricultural circles there is recognition that biocide use needs to be minimised, as chemicals used are soon superceded by a naturally produced 'counter chemical'. In MEDCs strict regulations apply to the use and application of biocides; with resulting increased use of biotoxins or disease resistance being built into the genetic structure of plants.

Amounts of biocide used/year (in tonnes) (INCREASE ↑)

Country	Tonnes
USA	380 000
Italy	174 000
Brazil	68 000
Spain	45 000
Mexico	34 000
UK	30 000

[Source: FAO]

7.4 Future trends

After studying this section you should be able to understand that:

- *concerns about the environment and animal welfare have led to an increased interest in more sustainable agriculture*

LEARNING SUMMARY

Sustainable agriculture

EDEXCEL A	Some U4
EDEXCEL B	U5
OCR A	U5
OCR B	U5
NICCEA	Some U5
WJEC	U5

Within the EU by 2000 it is expected that 10% of countries will be running sustainable operations.

Sustainability is really the millennium buzzword. Know and understand it!

Sustainable farming conserves nutrients, soil and water, whilst negating the need for lots of fertilisers and biocides. Sustainable farming is unable to compete with OIA in its intensity of yield, but competes instead through its appeal to its customers. Such individuals are concerned about nutritional quality, health and taste. Further, farmers are increasingly concerned about what they are doing to the environment, causing soil erosion, water contamination and so on. Sustainability in agriculture can range in the MEDC from using a low tillage plough disc, which affects only the top layer of soil, to fully fledged and accredited small-scale organic farms, that may well use no-tillage agriculture or perennial polyculture. In the LEDCs sustainable agriculture usually involves appropriate or transfer technologies and invariably returns the country back into subsistence management of farms.

Back-yard gardening!

Some 12.5 million hectares worldwide are successfully farmed in a sustainable way. Yields have doubled and tripled, and are continuing to increase far in excess of anything GM crops have to offer. Support for this is growing among farmers, trade unions, consumers, indigenous peoples and diverse public interest organisations. Not, however, from industry, because when farmers are free to keep their seeds, and to harvest and sell their produce locally, corporate monopolies cannot hold the hungry to ransom.

CASE STUDY

Sample question and model answer

The theory put forward by Von Thünen to assess agricultural location is assessed in this question.

1

Study the table below which is based on the ideas of Von Thünen.

LR – Y(M – C –Td)

Where: LR – locational (economic rent);
Y – Yield of a crop (tonnes per hectare);
M – market price (per tonne);
T – transport cost (per tonne);
d – distance from the market (kilometres).

With this type of question you'd be penalised if you missed off the units.

C = production costs/tonne or /hectare.

(a) (i) Complete the key to the formula by defining C. [1]

(ii) State *three* assumptions made in the Von Thünen model. [3]

Care! Bullet points should be avoided. Write in prose whenever possible. The candidate 'gets away with it here'!

- That there is just one form of transport.
- Transport costs are proportional to distance travelled.
- A uniform, isotrophic plain exists.

(b) Study Table 1 below.

	Y	M	T	C
	(tonnes)	(£)	(£)	(£)
Wheat	4	50	2	2
Potatoes	10	200	5	5

Table 1

Use the data from Table 1 to calculate:

(i) the locational rent for wheat at 10 km from the market; [2]

Always include working. It asks for it here!

Space for calculation

Always take all your equipment to your exam room. Here a calculator might be useful!!

LR = 4 (50 – 2 – [2 x 10])
LR = 4 (50 – 22)
LR = 112

(ii) which of the two crops would be grown at 5 km from the market. [3]

LR wheat = 4 (50 – 2 – [2 x 5])
LR = 152
LR potatoes = 10 (200 – 5 – [5 x 5])
LR = 1700

You convince the examiner you know what you are doing by adding this calculation.

CONCLUSION: potatoes are grown 5 km from the market.

Sample question and model answer (continued)

(c) (i) Suggest why transport costs are higher for potatoes than they are for wheat. [2]

Potatoes are reasonably delicate and need to be handled with care, which means the handling methods used to collect and store and transport grain are unsuitable.

Some slight irrelevance here – but the candidate does draw out the right point.

(ii) Suggest why today distance from the market is no longer such an important influence on the pattern of crop production as it was at the time of Von Thünen's analysis. [3]

The transportation in Von Thünen's time was not as good as it is today. Road transport today is fast, can be refrigerated, etc. All in all transport is more economic and efficient these days.

Not bad – has the idea of 'new' transport methods – less/nothing on bulk handling however!

(d) Comment on the economic and environmental problems associated with wheat production at the economic limit of cultivation. [6]

When two areas require 'comment' it is very important that balance is achieved or marks will be lost.

Economic production of wheat usually involves monoculture, this is extremely damaging to the environment leading to poor soil conditions and soil erosion, as the soil's nutrients and soil water are taken up and used (and not returned) to the soil. In economic terms, the huge investment in machinery and so on really does leave the 'market' vulnerable when prices drop.

Though it doesn't flow particularly well the critical points are in place.

Comment

This was a competent answer. One which shows depth of understanding of the issues related to farming. But also the candidate confidently attempts the supporting maths necessary in this question.

London Exams 'A' specimen papers

Practice examination question

Increases in agriculture productivity in the last 30 years have imposed an unacceptable cost on the physical environment. Discuss with reference to the EU.

OCR specimen paper

Chapter 8

Synoptic assessment

What is synoptic assessment?

To study a topic synoptically is to 'afford a comprehensive mental view', i.e. to consider all perspectives and factors contributing to the complexity of reality. **Synoptic assessment**, therefore, is 'a form of assessment which tests candidates' understanding of connections between different elements of a subject'.

How will the synoptic units have been taught to me?

Synoptic assessment is part of all A2 specifications. Your teachers may have integrated synoptic elements and themes, linked and inter-related topics in the studies and case studies that you've covered; or you may have been taught a range of synoptic issues as a discrete unit or section of work, probably near the end of your A2 course. There are advantages and disadvantages with either route.

What are the general areas that are relevant to my synoptic studies?

This was a requirement imposed on all exam boards by QCA, the group that approve specifications.

All geography specifications require knowledge, understanding and skills related to:

- the fact that geographical processes all have multiple interactions
- the fact that the environment needs to be stewarded to ensure a sustainable future
- the fact that a geographic understanding helps and assists with decision making and planning, and can also assist with hazard prediction
- the fact that governments, organisations and communities can determine geographical outcomes.

How do I ensure success in synoptic papers?

- All questions will want you to demonstrate your understanding of geography in a discursive way. According to the Oxford English Dictionary, discursive means 'to ramble or not stick to the main point'. Do ensure that your synoptic discussions do have a focus and do follow a thread! In other words don't allow your thoughts and ideas to wander too far from the thrust of the question as set!
- Do remember that you may be required to apply your knowledge and understanding of the geography that you've been taught to unfamiliar situations.
- Crucial to your success in answering questions on synoptic papers in general, and in the more unfamiliar situations outlined above, will be your ability to draw on a wide range of case studies. You would be expected to be aware of recent (i.e. during the period of your AS and A2 studies) and the more important of the familiar older/well-used studies.
- As you prepare for the synoptic paper ensure that you have a range of studies that cover many scales, locations and stages of development.

Read the reputable press; or use the internet regularly.

How can I prepare synoptic issues for the examination?

In the actual exam you may well list these links in your essay plan. Then cross them through as they are included in your essay!

Each issue in the synoptic section that follows in this book is set out in the same way. An outline of the issue, topic or problem is offered and then there is an amplification or a detailed case study offered in an attempt to address the issue. Each topic is then summarised. Possible general synoptic links are also established.

Just fifteen issues are covered here, they cover and reflect the most common denominators between the eight AS/A2 specifications. Those same eight specifications suggest some 108 issues that might be covered; you need to be aware of how the issues covered here fit into your specification and what additional issues/topics you need to cover.

In summary

- The synoptic sections of all the specifications allow those who are well prepared to score freely and highly. Do remember that a good 'score' here could well determine your final grading.
- You must know a range of geographical subject matter and be able to connect it all together.
- Finally, don't forget the importance of the human perspective in all the themes and issues you explore.

Issues 1–3

1 Climate change and river flooding in Asia

Most hydrologists and the techniques they have established to measure floods see them as random events in a stationary sense. However, river flooding is increasingly teleconnectable, a climatic event linked to climatic anomalies.

Amplification

Floods are a normal occurrence for the people of south Asia. Every year, floods destroy crops and displace the region's inhabitants. Floods are most common in the lowlands of Bangladesh where the Ganges and Brahmaputra rivers meet. Floods are both a hazard and an asset. While the floods are drowning crops and livestock, and damaging property, they are also fertilising the soil in the form of dissolved and suspended solids.

Projection and observing

Global warming has the effect of increasing temperature difference between the land and sea surface of the Indian subcontinent! The higher winds increase evaporation resulting in the transport of huge quantities of moist air. This leads to a greater number of rainy days during the summer monsoon with 20% – 25% increased rainfall. Observation and research reports show significantly increased trends of extreme rainfall over Bangladesh, West and Central India, Pakistan, Sri Lanka and China: increases in the order of 5% to 18% have been recorded since 1970. The result of all of this has been an increase in abnormally high floods. In the last decade in 1991, 1993, 1997, 1999 and 2000, disastrously huge floods have inundated Asia. Other 'global warming linked' phenomena have also contributed to this pluvial increase, El Niño is one of them, as is increased melting of the Himalayan snow pack. Both affect the magnitude and intensity of the monsoon in the area.

CASE STUDY

Hyderabad

India, late August 2000, flooding from the heaviest rainfall in 50 years, with origins in a fierce summer cyclone, killed 2000-plus residents and destroyed 10 000 homes. At least 150 000 were effectively displaced with winds approaching 200 kph. India's most fertile rice growing regions had also been destroyed, some 178 000 ha in total. This storm followed in the wake of a huge storm that killed some 350 people earlier in the month.

Most fatalities in Hyderabad seem to have fallen victim to poor house construction, as mud-walled houses collapsed around them.

Some 20 cm of rainfall was received in about 24 hours, one third of the city area's annual rainfall receipt.

The floods in India followed catastrophic monsoon rainfall in north and north east India, Bhutan, Nepal and Bangladesh. The floods killed several hundred and left one million homeless. Damage in Hyderabad, India's second city, will cost billions of rupees to put right!

CASE STUDY

Rebuilding may well necessitate a rethink over building styles, as brick and stone buildings were the only ones to survive the deluge.

At the same time!

In India's north west province of Gansu, 19 people died and 24 were injured.

This latest flooding occurred when high water poured through five cities in Gansu. The water flattened more than 1200 homes and damaged 5520 ha of wheat and other crops. 30 bridges and many water retention facilities were destroyed.

CASE STUDY

Summary

Possible synoptic links with:

- Rivers
- Settlements
- Development
- Climate

It has been scientifically proven that as CO_2 concentrations increase so mean annual global precipitation will increase in the order 3% to 15% (by 2070). Consensus suggests that greater amounts, 20% plus, may well fall in areas affected by monsoon rainfall. Heavy rainfall events have increased markedly in their frequency and intensity in the period 1961–2000 in the monsoon region. So if an enhanced greenhouse effect is forcing hydrological cycle changes what future is there for the hundreds of millions of coastal lowland/estuarine dwellers of Asia? Certainly into the new decade much will have to be done to ensure the viability and permanency of the built-up areas. There will be much human intervention in this area to ensure the area's long term future.

2 The environmental threats of climatic change

Introduction

1998 was the hottest on record. The record drought year. The deadliest hurricane of recent times (Mitch) and most hurricane events (33). El Niño was the strongest of the century.

The course of climate change has its origins in industry and transport pollution. Climate change is leading to rising sea levels, species extinction and increasingly dramatic changes in extreme weather. The frequency and severity of phenomena such as extremes of heat and drought, intense precipitation, floods, hurricanes and El Niño events has increased. With every fresh extreme weather event there is also an economic cost to pay. It has been estimated that in 1998 there were 44 000 deaths and $89.5 billion in economic loss around the world. In the last 20 years the number of disaster declarations has doubled every year.

The extremes of Florida's weather in 1998

'At various times through 1998 Florida's residents prayed for rain and prayed for it to stop.'

Florida's southern farmers estimate floods cost them $150 million in crop damage; whilst droughts cost Florida $150 million in losses over a range of crops. Floods also affected residential areas. Violent tornadoes destroyed and

CASE STUDY

damaged homes in and around Miami. During the drought, 'wildfires' destroyed half a million acres of land and 400 homes. And then came the hurricanes, Georges in late September and then the biggest of the lot, Mitch.

Florida's six weather-related disasters; hurricanes, tornadoes, heavy rain, drought, fire and flood were the most of any state in the USA. 'To have had any one of the events would not have been unusual on its own. But having to face six in one year is very different.'

Source: Naples Daily News, 27.12.98

CASE STUDY

Summary

The 'cure-all'?

The United States and 174 other countries have ratified the United Nation Framework Convention on Climate Change, the global warming treaty was signed at the Earth Summit in 1992. This treaty commits parties to implement programmes to reduce their greenhouse gas emissions with the objective of stabilising these emissions at a level that prevents dangerous interference with the Earth's climate.

Recognising that little progress toward this objective had been made, parties to the global warming treaty negotiated the Kyoto Protocol in December of 1997. This protocol will bind developed countries to reduce greenhouse gas pollution emissions to an average of 5.2% below what they were in 1990.

The first steps have not come easily. International negotiations have been slowed by a powerful coalition of electric producers, car manufacturer and oil and coal companies who want to continue to emit global pollution at current, unrestricted levels. In fact President Bush, March 2001, announced the USA's intention not to ratify the protocol.

The cost in money, lives and to property that global warming will inflict on us if we do not take steps to reduce polluting emissions is immeasurable.

Possible synoptic links with:

- Industry
- Development
- Climate
- Atmosphere

Protecting countries and communities from global warming induced weather destruction should be high on the agenda of all MEDCs to achieve these reductions.

- Increased fuel economy standards for cars are a must.
- There shall be set standards for power plant CO_2 emissions.
- And there should be increased support for companies and individuals who are energy efficient and use renewable energy.

3 Disaster and response – lessons from Rostaq

Recent years have seen the emergence of disaster management systems, particularly when there are major humanitarian crises. It has been recognised that single organisations can no longer make adequate responses.

Rostaq Earthquake, Afghanistan, 4 February 1998

Details 2223 people killed out of a population of 17 600 in a quake of 6.1 on the Richter scale

The area Remote and mountainous

The people Agro-pastoralists involved in an ongoing civil war

Difficulties post-quake Security for those offering humanitarian aid, distance from main settlements, lack of vehicles and fuel, poor weather

CASE STUDY

Goals post-quake To eliminate threats to life and health, return displaced people to their villages

Time scales 2 days after the quake first help arrived!
2–12 days after the quake medical emergencies are landed
13–30 days after the quake the main emergency food and non-food arrives
81 days after the quake NGOs depart Rostaq.

Pressures

Less than a week after the quake western journalists were in place in Rostaq. Their presence created enormous pressure on the agencies to be seen to be doing something, and quickly.

Results some trucks and supplies get through in the early phases of the UN response, but a week after the quake the mission is paralysed. Renewed efforts eventually secure airdrops and large donkey caravans to get the food and aid to those that need it.

CASE STUDY

What has been learnt from Rostaq?

The response by various NGOs at Rostaq were such that brief windows of opportunity immediately following the quake were lost and that the logistics of multi-agency responses have to be co-ordinated. From Rostaq much has been taken and many lessons have been learned.

- The temporary alliance amongst NGOs and more major players must be centralised as situations and responses get paralleled. A waste of resources results.
- Trade-offs have to be avoided. They cost lives, some 700 estimated in Rostaq, as decisions about decision making, speed, reliability and cost responses were made.
- In effect, lower effectiveness occurred during the early life-saving phase of the rescue and higher effectiveness was experienced during the later 'quality-of-life' window.
- Co-ordinated and controlled large groups of NGOs can work well and complement one another and, importantly, save lives.

Postscript

Possible synoptic links with:

- Hazard management
- Population
- Settlement
- Development and human welfare

On 2 May, another, even worse earthquake struck the same region. This time the international agencies responded in more co-ordinated ways. Helicopters were quickly procured for needs assessments and relief deliveries; the Swiss Disaster Corps was brought in to connect all activity centres with telecommunication; and Islamabad was made the command centre for both the UN and other NGOs. The relief community had learned some lessons from its earlier experience.

Sample extended writing question and answer

1

Hazards are becoming more frequent. Discuss

What figures? State some.

According to UN figures the number of hazards per year is increasing. The 1990s were the worst decade ever for disasters with 1999 the worst on record for environmental tragedies. There are two possible reasons for this, the first is that the number of hazards and consequently disasters is increasing. The second possible reason is that people are populating hazard-prone areas at an increasing rate and as such natural hazards have become disasters. In this essay I hope to look at both these possibilities and to discover whether or not both of them are true.

Excellent point – the best students will cover this ground.

Good knowledge.

Principally in LEDCs.

Suggest ways of reducing the human factors.

Was it because the scheme failed?

Multiple hazard Japan – an ideal exemplar. But inadequately developed here.

The 1990s was marked out by the UN as a decade for disaster reduction. The UN proposed to increase people's awareness of hazards and therefore help them to survive. It is the presence of humans that turn a hazard into a disaster and the UN wanted to reduce the human factor and therefore reduce the number of disasters. Unfortunately the scheme failed and the 1990s was the worst recorded decade for disasters. For many people it is a necessity to live in a potential hazard because of limited resources and a growing population. During the 1990s the world's population grew at about 10 000 people per hour putting an increasing pressure on resources. Nowhere is this more apparent than in Japan, an island in constant threat of a major earthquake. People in Tokyo live with the knowledge that an earthquake could destroy their whole city.

Much better use of examples.

Disasters are therefore increasing because people are increasingly living in hazard areas. 500 million people live within range of a volcanic eruption. The majority of people in the world live in areas where flooding is a threat. Recent floods such as in Venezuela with over 10 000 dead and Vietnam where 90% of the Dong Thap province is under water and the province capital Cao Lanh under 1 m of water. People living in hazard areas is happening more and more; in LEDCs particularly in cities where people live in shanty towns on unstable slopes. Landslides caused by this kill many people.

Good paragraphing shown here ensures pertinent parts are kept together.

Good, brings the study full cycle.

Good link and knows the terminology of the subject.

Good example.

The second reason for the increase in disasters is that the number of hazards are increasing. This is caused by changes in world climate, particularly global warming. This has the effect of raising sea levels thus further reducing the land for people to live on and creating a flooding hazard of its own. The warmer weather will increase evaporation and because of the dynamic equilibrium of the hydrological cycle, this means more precipitation. This will increase flooding and landslides. The increased temperatures will create more powerful cyclones and hurricanes. In October of 1999 the Orissa cyclone killed 10 000 people in India. Low-lying countries like Bangladesh will suffer the most from this. Atlantic hurricanes are now 40% more intense than 30 years ago. Conversely, many dry areas will become more arid and this will create its own problems. If global warming continues, these hazards and disasters can only become more devastating.

Super – a point that will be missed by many!

A potential third explanation for the increase in hazards and disasters is because of improving communications across the world. Many more disasters are being detected and information about them is spread more efficiently and quickly.

The number of hazards and disasters are increasing, hazards due to climatic change are disasters mainly because of the expansion and growth of the human population. Natural disasters kill more people than war does and like

Sample extended writing question and answer (continued)

Sound evaluative conclusion, that draws many threads together.

war the effects of a disaster are felt for years afterwards. People are left homeless, without food or medical care and without hope. If the number of disasters is to increase then more must be done to make people aware of the hazards and more must be done, mainly in LEDCs, to make sure that these disasters are survivable and their effects are limited.

Comment

After a rather dodgy start, this student managed to turn the whole essay around. With some rather knowledgeable and well researched and conveyed ideas. This was a good synoptic essay drawing on a number of areas of the specification. It was both evaluative and thorough.

Practice examination question

1

(a) Draw an annotated diagram to show the Greenhouse Effect. [7]

(b) Is global warming a natural phenomenon or largely a result of human activity? [12]

(c) Why should scientists need to 'model' the consequences? [6]

EDEXCEL B specimens

Issues 4–7

4 The water crisis – sustaining water, easing scarcity

80 countries with 40% of world population don't have enough water.

It is hoped here to give an insight into how water stress/scarcity and how natural thresholds relating to population–environment interactions operate. It is well known that expanding populations and economies place increasing demands on water supplies and that the problems are becoming more acute. To date the water supply has always expanded to meet population requirements. For the time being there is plenty to go around, nearly three-quarters of the earth's surface is water, but the provision of water supplies is chaotic and piecemeal. There can be intense competition for water resources, especially in the LEDCs. In the MEDCs water has become a tradable commodity; federal and local governments have to provide it at the right price, quality and quantity to be attractive to investors etc. There is a real risk around the globe that serious social and political conflicts will arise because of water shortages.

Amplification

CASE STUDY

The Case of the Tigris-Euphrates Basin

A situation with serious international implications is the demand for the waters of the Euphrates by Turkey, Syria and Iraq. The Euphrates is the primary water source for millions of people who depend on it for power generation and irrigation in an extremely arid climate. Conflict over water in this area is decades old. It has intensified in recent years as a result of a massive Turkish dam building programme known as the Greater Anatolia Project (GAP), designed to provide a supply of water and power adequate to fuel the development needs of Turkey's population, which is growing at 1.6% annually. When completed, it will provide Turkey with a generating capacity of 7500 megawatts of electricity-nearly four times the capacity of Hoover Dam – and open up at least 1.5 million hectares of land to irrigated cultivation.

Full implementation of the GAP system of dams could result in a 40% reduction of the Euphrates' flow into Syria and an 80% reduction of flow into Iraq. This will reduce the electrical output of Syria's Tabqa Dam by up to 12% of capacity, while Iraq could lose irrigation water to one million hectares. The levels of salinity will increase as well as the amounts of agricultural and industrial pollution in the remaining water being conveyed into Syria and Iraq.

Syria and Iraq have already threatened war over their access to the Euphrates. As the populations of these nations continue to expand, driven by fertility rates well above the global average, the competition for fresh water between agriculture and development could endanger stability in the region.

The Nile

In Egypt they have recognised for some time that the headwaters of the Nile might be developed, it too has threatened war, to preserve its access to fresh water. Ethiopia has already built some 200 small dams since emerging from civil war and famine.

The Edwards Aquifer – South Central Texas, USA

Described as one of the USA's 'most wondrous aquifers', because of its storage capacity, flow characteristics, water producing capabilities and efficient recharge ability, the groundwater from the limestone of the

Edwards Aquifer is designated as the 'sole source' of drinking water for the 1.5 million people of San Antonio and the Austin–San Antonio corridor. It is also vital to agriculture, the light industry of the region, tourism and recreation. Growth in all areas is posing a major problem because of over drafting. This has resulted in the implementation and enforcement of critical management plans and water conservation ordnances, and led to an increasing number of water-user disputes. In an attempt to defuse the growing number of disputes the San Antonio Water System and ALCOA (the Aluminum Company of America) have joined forces to pump water from a site 100 miles NE of San Antonio. The pumping will break San Antonio's addiction to increasingly scarce local groundwater.

The proposed new water source might well safeguard San Antonio's future; but what of the area around Bastrop County where the water is to be sourced? Texas State law suggests that San Antonio Water and ALCOA would have to 'replace and repair' wells damaged by pumping, even though landowners in Texas can pump from their own land, without limit, whether this dries up the aquifer or not.

CASE STUDY

Groundwater is subsurface water that fills the pore spaces of soil and rock below the water table. It is in constant motion.

Summary

The whole world watches disputes in the LEDCs and MEDCs of the world with concerned interest. The then UN Secretary General Kofi Annan, in 1997, stressed that water 'was the most urgent issue on the global agenda'. The projected world population growth will ensure that over the next couple of decades there must be cooperative interest in shared water resources. The USA has gone as far as suggesting water-rich countries negotiating trans-boundary solutions, moving water to areas of scarcity. However, the challenge of reconciling competing demands and claims to water will occupy many governments for the time being.

Possible synoptic links with:

- Population
- Tourism growth
- Rivers
- Hydropolitics
- Sustainability

5 Disputes over the resources of the oceans

Why argue over the sea?

There has been an increasing interest in oceanic resources in recent years; this is to do with the realisation that the earth's stock of physical resources are being depleted at an increasingly accelerated rate. Covering nearly 75% of the planet, the oceans embrace most of the earth's biosphere – the region where life occurs. The ocean's resources are a treasure for current and future generations. The value of these resources is inestimable; the value of indirect resources (for carbon storage, atmospheric gas regulation, nutrient cycling and waste treatment) for instance, is put at $11.7 trillion alone!

As population increases, demands for food, products and services from the oceans and seas will increase. Likewise demands for living and recreational space along the shoreline will also grow. This has led to a scramble by nation states and those interested solely in commercial exploitation to stake claims and to appropriate sea resources. Conflicts and disputes over claims for/to sea space, fisheries and minerals around the world have proliferated. There are about ten major disputes/year for the International Court of Justice to legislate upon, and the numbers of such cases are increasing, e.g. Cameroon–Nigeria, are disputing oil reserves and exploration off the Bakassi Peninsula; Spain–Canada, are disputing fishery limits off Canada; Japan–Korea/Japan–China, are contesting island ownership off their shores.

Amplification

What are the sea's resources?

Navigation space

There is a basic right of unimpeded passage and navigation, this established through the Law of the Sea Conferences. There are at present some 55 000 ships plying the seas, about 35% are oil tankers. The most congested points are the straits and ports.

Living resources

At present the oceans of the world contribute 14% of animal protein for human consumption. This could substantially increase in the future. Clearly stocks will have to be managed but, it's been estimated that up to 110 million tonnes of fish/year could be sustainably landed in the future. High protein krill and fish farming will also contribute too.

Mineral resources

- On the continental shelf, i.e. sand, gravel, tin, diamonds, manganese nodules, phosphate nodules and minerals of all kinds.
- Under the continental shelf, i.e. oil, gas/coal and sulphur.

The sea as a waste sink?

The sea has a great capacity to dilute and degrade waste. It negates the installation of expensive pollution control on land. In the future this will have to be controlled.

Wave energy and tidal power.

Both are of strategic and economic use!

CASE STUDY

It should be clear from this brief coverage why there is this rush for marine space and why the seeds of future conflict exist, and will continue to be sown. The case studies below highlight two issues to do with the sea, those of conflict and stewardship.

The muddle that is the South China Sea

China has again imposed a two-month ban on fishing in the South China Sea. The Philippines fear the ban could bolster China's sovereignty claims in the area. The ban is the second to be imposed by China in the South China Sea. Philippine diplomats have said that while the ban could mean less Chinese presence and 'intrusions' in the Philippine-claimed Kalayaan islands and Scarborough Shoal – all located in the South China Sea – it could also bolster Chinese authority over the Philippine-claimed territory. Both China and the Philippines claim the shoal. The Kalayaan Islands are a part of the Spratleys but the shoal is not. The Spratleys are claimed in part or as a whole by China, Brunei, Malaysia, the Philippines, Taiwan and Vietnam, with China claiming the whole South China Sea.

Source: The Guardian April 2000

Oil exploration around the Falklands

After drilling the first six exploration wells in the North Falkland Basin, oil companies have found at least two hydrocarbon systems. The search for commercial accumulations will be the next major stage, but timing may depend on improved oil prices. The operators, Amerada Hess, Shell, LASMO and IPC form a consortium called FOSA (the Falklands Oil Sharing Agreement) who are responsible both for the exploration of the oil but also for the careful environmental extraction of it. The Falkland Island Government is paying close attention to safeguarding the environment in the event of a major commercial oil discovery. The Falklands Island Government and the oil companies working in the area are anxious to maintain the pristine conditions that exist in the region for future generations.

Source: The Guardian April 2000

CASE STUDY

Possible synoptic links with:

- Coasts
- Development and human welfare
- Population and resources
- Energy

Summary

Whether the sea can solve the problems of world resource depletion rests very much on whether ways can be found to reconcile the uses that will be made of the sea, and the interests of both coastal and landlocked states (such states have a right to the benefits accrued from common property resources too!). Additionally, usage of the oceans must be balanced with appropriate stewardship.

6 Drought and its teleconnection to ENSO (El Niño Southern Oscillation)

Teleconnections are atmospheric reactions between widely separated regions.

ENSO occurrences are global climatic events linked to various climatic anomalies. Drought in particular has a strong teleconnection with ENSO.

The mechanism

The press is full of El Niño references – keep abreast of them.

Every two to seven years ocean currents and winds shift off the coast of western South America bringing warm water westward. In recent decades, scientists have recognised that El Niño, as it is called, is linked with shifts in global weather patterns. Its effects last between 14 and 22 months. The southern oscillation, 'a seesaw of atmospheric pressure between the E Pacific and Indo-Australia', is closely linked to El Niño. Generally El Niño refers to the oceanic properties and southern oscillation refers to the atmospheric component, see diagrams below:

Figure 8.1 *Normal circulation: **with** southern oscillation*

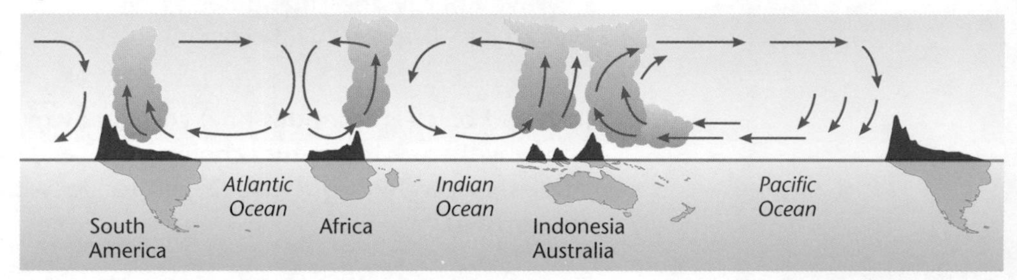

*Figure 8.2 ENSO event: **no** southern oscillation*

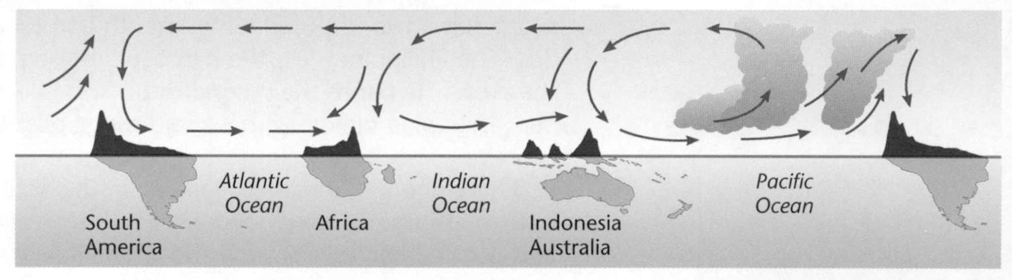

Source: Bryant, Natural Hazards 1991

Amplification

Drought is defined as an extended period of rainfall deficit during which biomass is curtailed.

Researchers have found that during an ENSO event drought can occur virtually anywhere in the world. Central America and various parts of the USA, the islands of the Pacific Basin and India all show strong relationships. E Australia, for example, has been subjected to abnormally dry conditions for much of the late 1990s, these dry conditions have decimated crops and caused bush fires (on a year to year basis since 1997 drought has cost Australia A$2 billion annually).

CASE STUDY

1998 – 'the hottest and driest year since 1866'

Impacts around the globe show how drought affected countries with differing levels of development.

Florida Dry weather and record heat; water restrictions in areas (e.g. Delton placing restrictions on outdoor water use such as washing cars and watering lawns); over 3600 ha burned in state-wide fires; the state banned all open burning; fire restrictions were put in force in national forests; firefighting teams were brought in from Georgia, Arkansas and Alabama.

New Zealand. Summer drought and economic problems from the Asian crisis; 1 in 100 year drought in the Wairarapa/Tararua area; 7.1% unemployment in the first quarter of 1998 (129 000 people – up 7 000 from the previous quarter and 13 000 more than a year ago); drought affected farming and ranching activities.

Indonesia Drought and economic crises (although, recent rains have eased the situation); 40 million people live under chronic marginal circumstances (8 million people at risk of food shortage); A$50 million in humanitarian aid (grain, medical supplies, feeding and agricultural programmes, drought co-ordination, and employment programmes) and a $1 billion loan fund was made available by Australia.

Nepal Malnutrition and diseases (influenza, cold, fever, dysentery, and diarrhoea) in the Humla district brought on by poor harvest now caused by drought (some of the diseases have now been brought under control).

Namibia Drought in the north east; 25 000 people needed food aid in Caprivi and Kavango regions (these families did not plant in 1998 because of drought conditions).

CASE STUDY

Summary

Possible synoptic links with:

- Atmosphere
- Development and welfare
- Climatic change

Being able to forecast and anticipate droughts (they are termed slow onset hazards) would be useful as the 100 or so countries that continue to have long periods of drought experience severe consequences to their social and economic development, because of this hazard (see figure 8.3).

Figure 8.3 *Social consequences of drought*

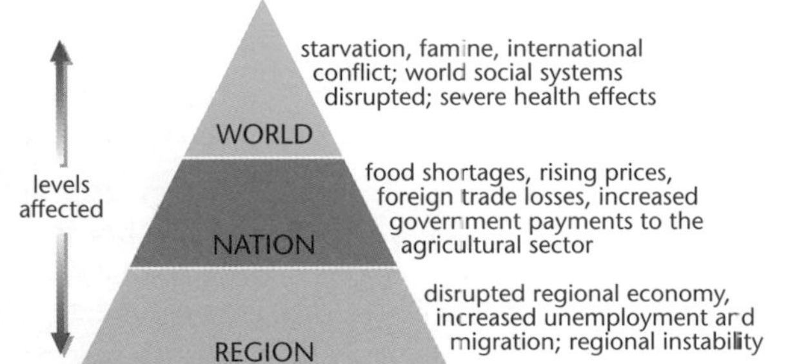

7 Saving the world's soils

Nature of the problem

Soil erosion is just one form of soil degradation. Others are soil compaction, low organic matter, loss of structure, salinisation and acidity problems.

In many places around the world the natural equilibrium of the soil is being disturbed by human activities. Many people view the land as a resource that belongs to them, to be used in whichever way they think fit. This is an abusive way of using the soil that will eventually destroy the land we seek to use. Across the world, farmers are attempting to feed 90 million extra mouths per year. This is against a backdrop of massive soil erosion, of some 24 billion tonnes/year. Cropable land is actually shrinking!

Amplification

Under normal conditions soil erosion is a natural process occurring under non-agricultural conditions; i.e. the removal of material by erosion at the surface and the addition of material by weathering is part of the fine balance typical of natural

Sheet, rill and gully erosion are the main forms of water erosion. Wind erosion is able to strip as much soil as these three processes put together.

soils. Soil erosion only becomes a problem when the rate of removal of soil by water and/or wind exceeds the rate of soil formation. Accelerated soil erosion normally results from the removal of the surface protection afforded by natural vegetation, but can be the result of changes in soil conditions brought about by climatic change. An explanation of the causal factors of soil erosion is shown in the diagram below.

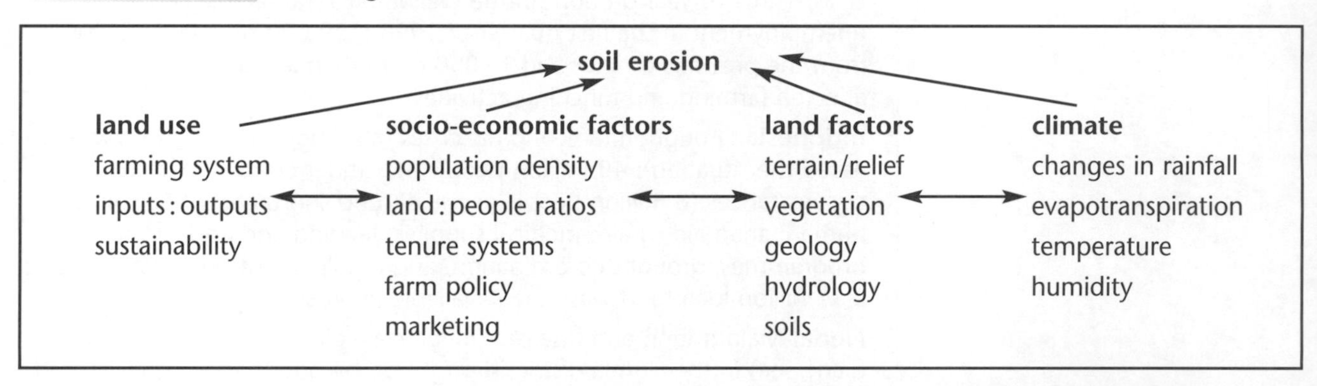

CASE STUDY

War, poverty and disease? Africa's soil may be its worst enemy

'Deforestation, overgrazing and harmful irrigation practices are transforming vast areas of the world's poorest continent into virtual wasteland. National economies are crippled, fueling social and political havoc.'

Farmers are feeling the effects of generations of exploitation. Africa was once a place of plenty, hillsides now lie bare from years of erosion. Unless current trends are reversed, Africa will be unable to feed two-thirds of a projected 1 billion population in 2025. Already 200 million people are chronically malnourished, double the figure 30 years ago. Most Africans today eat about four-fifths of what they did in the 1950s, when the continent was a net exporter of food. And it all comes down to the soil.

In an era when chemical fertilizers, pesticides and modern machinery have boosted food production to record levels in the United States and Canada, African farmers struggle to survive using shortsighted farming methods that often degrade the fragile soil. For generations, subsistence planters, stock herders and loggers have engaged in a battle with the land. They slash and burn virgin forests and savanna grasslands, often replacing them with cash crops that leech life-giving nutrients such as potassium and nitrogen from the soil. Rapid population growth in burgeoning urban slums has increased the pressure on the overtaxed soil. UN studies suggest 'the low fertility of African soils is the single most critical impediment to the region's economic development'. 'Real progress will not be made in Africa until the problem of degraded soil is addressed.' Already 2.2 million km^2 is classified as degraded land.

As long as civil wars and political upheaval are rife, with foreign aid money going first toward peace-keeping missions and refugee disasters, the agricultural problem on the land and development will be neglected.

Source: Associated Press 16.10.99

The problem of wind erosion in North America

Wind erosion is a serious problem in many parts of the world. In the USA, 70 years after the notorious black blizzards of the Dust Bowl, the Great Plains are still losing 30 million ha of land to damage from the wind per year. Blowing soil reduces seedling survival and growth, depresses crop yields, lowers the marketability of vegetable crops, makes crops susceptible to disease and contributes to pathogen transmission. Put together this amounts, in Kansas, to a 340 000 bushel yield reduction of wheat and a 540 000 bushel yield reduction of sorghum. Most of the eroded soil enters the atmospheric dust load, where it pollutes the air, fills ditches, fouls machinery and imperils human and animal health, e.g. on a 500 m stretch of Kansas highway 965 tonnes of soil was removed in 1996! Wind erosion in the USA is a threat to the sustainability of the land as well as the quality and viability of life for rural and urban communities.

Source: US Wind Erosion Research Unit

Summary

Management of soil degradation at a global, regional and local scale is a complex issue; one which represents one of our most challenging environmental problems. The challenge according to the World Soil Charter (part of FAO) 'is for governments and landowners to manage the land for long term advantage rather than for short term expediency'.

***Figure 8.4** Spiral of poverty*

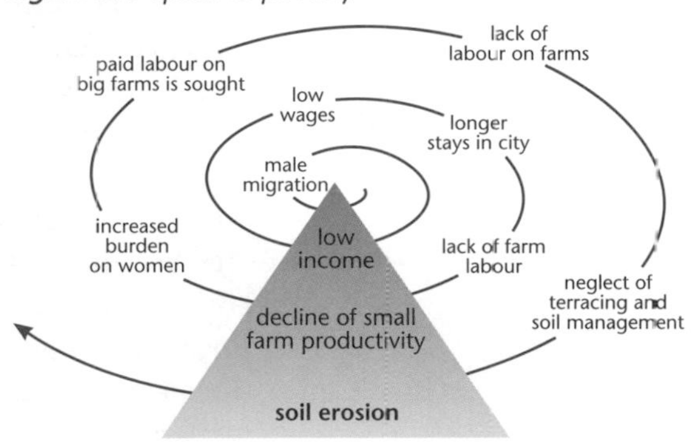

Possible synoptic links with:

- Soils and ecosystems
- Development and human welfare
- Climatic change and atmosphere
- Population

Sample extended question and model answer

1

Outline the demands placed upon and uses made of rivers due to population and settlement growth in fluvial environment.

Useful way into this short account.

This answer probably needs at least three references to uses and demands on rivers. This was achieved here.

Increased population and settlement growth can place strain on any area, but human activity is especially relevant in fluvial environments, with the demands of man and his use of the river liable to upset the sustainable fluvial environment. Residential demands on a river are particularly crucial, with all housing needing running water, and the only source being the river. Reservoirs, canals and water treatment plants have to be built to support this demand, and can often upset the balance of the ecosystem. An urban area's need for water is not the only affect it has on the river, with tarmac and buildings impermeable, causing a shorter lagtime between rainfall and the time the water enters the river. The same effect can be created by the removal of vegetation for building work and agricultural purposes, with less interception for precipitation available. The removal of vegetation is not the only demand farming places upon the river, with water from the river also needed for irrigation. River banks are eroded by livestock and lack of bank vegetation. Farmer's fertilisers, especially the use of slurry, can also lead to a process called eutrophication, with the treatments washed from the field in the rainwater entering the river system, causing an imbalance in the fragile ecosystems.

Connections between different uses, demands/effects were established.

These references to actual places need to be further developed.

Settlements and industry also cause a certain amount of river pollution, with dirty water flushed back into the river after earlier being diverted from it. Sewage works do help to improve the water quality of the river, but it is often necessary for local authorities to build complex systems to allow water to be stored until levels drop, when it can then be released to dilute the unclean water released into the river. Such a system has been incorporated into a scheme on the River Exe.

As well as these demands, population growth also causes an increase in the number of leisure activities that take place on the river. These activities can place further strain on an ecosystem, with boats undermining river banks and causing erosion, and petrol motors further polluting the water. This happens wherever water and leisure meet, a notable example being the pollution of the Norfolk Broads by the large holiday boats hired by tourists.

Human interests and ecological interests often come into conflict in fluvial environments and this is often triggered by increased development or a sudden population growth. Authorities try to limit any damage that growth on a river valley may cause, but invariably detrimental effects to the environment and river systems do follow.

Comments

This was a good extended piece that included references to a wide range of demands and uses. Importantly it looked to demonstrate an understanding of population and settlement. The focus seemed to be on the river rather than the valley, which suits the question as presented.

Practice examination questions

1 What factors continue to limit the expansion of food production in some parts of the world? [15]

The synoptic links for this question are complex, but only because they are large in number!

These are predictable subjects, which are favourites with examiners. This type of question with so many obvious links should be carefully prepared. You must have good notes and a range of good examples. Surf the net!

You will answer this question well if you have a sound but basic knowledge of the 'link' areas.

Student help

Feasible ideas are:

- climatic links
- population growth
- soil links (erosion/salinisation)
- development issues: disease, literacy and poverty
- changing agriculture – technology and farm size
- hydrological links: adequacy of supply and extreme events
- political issues.

2 Assess the major environmental outcomes for those countries that have increased food production to satisfy global population growth. [10]

Attempt to establish the links and connections between the different geographical themes and case studies.

Remember to include with your global overview, via the synoptic links, plenty of acceptable regional and local examples.

Student help

Synoptic links are feasible with:

- agriculture: intensification v extensification
- soil issues: degradation in particular
- ecosystems: deforestation and biodiversity loss
- groundwater/hydrological links
- global changes to do with sourcing of food.

Issues 8–11

8 Our waste is out of control

Background

Only Greece has a worse record than ours related to waste!

Man's use of materials in the modern world is massively wasteful. Industry is inefficient, only 9% of marginally input raw materials are used, cars on average waste 90% of fuel in their tanks. And the average MEDC household throws away up to 600 kg per person, per year. In addition there is much hidden waste material.

Until the last decade there had been little or no attention paid to this problem, in fact 'one-time-use' products and cheap consumer goods have actually exacerbated the problem. With greater attention being paid to the environment through various initiatives and dictates, much more attention will have to be paid to waste. With reduced numbers of places to safely dump material in the UK it is becoming a particular issue for us. Sustainable waste movement schemes and integrated waste management has to be instituted.

Amplification

In the MEDCs the waste management policy has been a predictable one:

The dinosaur disposal method!

The usual option in the UK is to dump our waste materials into landfills. Nowadays these are carefully engineered. 83% of the UK waste ends up in landfills.

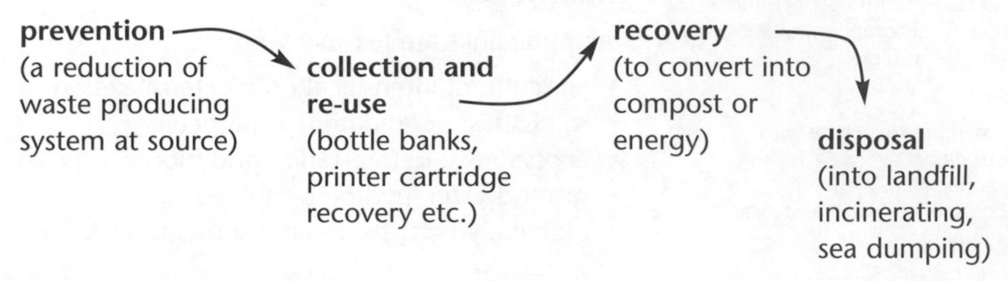

Landfill taxes have lead to a £10 million drop in official waste deposition.

Nature operates a closed system of recovery and recycling. Our current waste practices have actually caused a break in the system by creating billions of tonnes of non-biodegradable material that just won't breakdown naturally. Of the waste we actually produce only 5% is hazardous, some 60% will decay naturally, the remaining percentages are made up of dredging/sewage sludge and construction debris.

Eight million nappies are added to landfills/day.

52% of waste is recycled in Switzerland. 8% in the UK! (25% by 2005!)

CASE STUDY

The 'disposable' problem

Disposables have an after life! They represent about 4% of household waste, and fester and rot for up to two or three decades before they begin to breakdown. They use up to three times the amount of energy used to produce and clean terry nappies. Estimated costs to remove and dispose of nappy waste is in the order of £40 million.

Solution? The women's environment network has an alternative, arguing that terry nappies are more environmentally sound and cheaper.

Forget the take away – the throw-away's the problem

Every year Britain throws away up to 5 000 000 tonnes of edible food, valued at £400 million per year, with landfill tax and disposal costs at £50 million. This vast wastage troubles many, charities, the environmental groups and the major supermarkets. 20% of all food produced in this country is at present destroyed.

The principal driving force behind this dumping is due to government intervention to maintain fair prices! The 13 milion who live in poverty in this country at present only benefit from about 3000 tonnes of 'give aways'.

'Think globally, act locally'

Farmers' markets where growers sell direct are making a welcome reappearance in Britain. These range in size from high profile venues to the 'square' in the market towns of lowland Britain.

And why are these markets reappearing? Because:

- they reduce travelling
- they reduce unnecessary packaging
- farmers get the true value of their products

CASE STUDY

Summary

Possible synoptic links with:

- Sustainability
- Ecosystems
- Climate change
- Pollution

Regulating controls are probably the key to the problems of waste in this country (coupled with increased environmental incineration). Methods include: maintaining and extending the land fill tax scheme/educating rather than legislating/tax by weight all household waste (though this may encourage fly tipping)/encourage separation of waste/reduce bin size/encourage waste-free thinking.

9 Energy, the number one global issue!

Background

We all use energy in one form or another to cook, to keep warm, as a substitute for powering machines. These activities contribute to and signify economic development.

Energy also has another face. It is the face of environmental degradation. Environmental damage results when energy is used, due to the wastes that result. Some, such as coal, leave huge scars when they are extracted. They also contribute the majority of pollutants that affect the air. One consequence of fossil energy being burned is the release of greenhouse gases.

Non-fossil energy is not free from environmental consequences either. Nuclear power raises concerns about disposing of radioactive waste. Vast areas of Europe were affected when a fire occurred at the nuclear power station at Chernobyl in the Ukraine in 1986.

Other sources of energy also present environmental risks: hydroelectric dams flood the landscape, changing habitats, and altering river biosystems. Dams constructed in areas subject to earthquakes require engineering safeguards that greatly increase costs. Even windmills used to power electric generators stand out in the landscape changing greatly the natural appearance of large areas.

We rely upon energy in any discussion of economic issues surrounding the use of global resources, environmental issues, including atmospheric pollution and global change. Energy fuels the economies of countries. People respond to the benefits of energy use, rarely do they predict the negative consequences of increased energy consumption.

Two very basic questions must be asked when studying the geography of energy. First, where does energy come from at the global scale? Second, what purpose does energy serve us as individuals and groups of people?

Energy targets and harnessing the waves

The UK could generate all the electricity it needed from around its long coastline. However, there are enormous hurdles to overcome, not least being able to construct a viable generator. But the race to bring renewable energy to our homes and businesses is on: by 2010, 10% of electricity provided by the electricity supplier must come from such sources. It is thought that this requirement will help many small rural communities to initiate and run their own renewable power plants.

Islay in the Inner Hebrides may be one of the first communities to benefit; as they now have what is believed to be the world's first commercial wave-powered generator. No doubt other island outposts will watch this project with interest.

Part of the reason for the 2010 10% renewable energy requirement is bound up in the government's pledge in its report, *Energy – the changing climate*. That is by 2010 CO_2 levels will be cut by 20%.

The commission offered four ways in which these targets could be reached. Britain is embarking on a pathway that leads to a sustainable energy policy that protects the interests of our children and grandchildren and the generations after them.

Recklessly causing large-scale disruption to climate by burning fossil fuels will affect all countries. It is the poorest that would suffer most. We cannot expect other nations to do their part in countering this threat – least of all if they are much less wealthy – unless we demonstrate we are really serious about it.

The commission recognised the value of nuclear power in providing carbon-free energy but did not believe it was indispensable.

The report confirms that fossil fuel economies such as the UK's are on the wrong path, but it also shows that wind and solar power can break our addiction to oil, coal and gas. It makes clear that tinkering around the edges, which is what all governments are doing now, won't stop climate change wrecking our lives and economies in places like Mozambique, or prevent sea level rise flooding large parts of the UK.

CASE STUDY

Energy demand varies in different regions of the world. The greatest and most obvious variations are between the developed and developing countries. The greatest responsbility for reduction in energy use will fall on the industrialised countries. They are also the best equipped to develop alternative energy sources. The predicted curve for the industrialised countries would have to be reduced by the equivalent of approximately 30 million barrels of oil per day between 2000 and 2010, to permit a stable world demand, and to provide the developing countries with the energy they will need.

Summary

Energy is a resource that has many different forms and, depending on the kind of energy, it is distributed across the earth in a variety of geographical patterns.

The way energy is used affects the lives of people and makes it an issue of political, economic, and social importance. World primary energy resources, oil, wood, coal etc. are diminishing. Energy must be used wisely – both for the present and for the future.

Possible synoptic links with:

- Ecosystems
- Coasts
- Sustainability
- Settlement
- Population and resources

10 The precariousness of food supply

Background

It has been estimated that 35 000 people die of hunger every day and that one-eighth of the world's population is starving.

More than 40% of the world population also suffers from a lack of micronutrients such as vitamins or trace elements. Some two billion people are affected by iron deficiency. Around 1.6 billion live in areas with an endemic lack of iodine, and about 230 million children across the world suffer from Vitamin A deficiency.

Increased food production, which in the 1980s and 1990s kept pace with population growth, has in many countries fallen behind. But even in regions that produce sufficient food not all people have access to it because of their low income. The United Nations Food and Agriculture Organisation (FAO) in 1996 categorised 82 nations as Low Income Food Deficit Countries (LIFDCs). Half of them are in Africa. People who grow up suffering from chronic malnutrition have few chances in life. Their physical and mental development and their efficiency are limited. Damage to health suffered in early childhood is mostly irreversible.

Food as a basic need is enshrined in the Universal Declaration of Human Rights. The developed world has a duty to ensure food security measures are achieved:

Or does it?

- compensating for temporary food deficits due to unusual scarcity and/or a sudden lack of income or subsistence production
- compensating for chronic or structural food deficits due to factors with longer term impacts, and/or remedying an ongoing lack of access by population groups or households to sufficient food and food related services
- reducing and helping at/during the range of emergencies that cause food supply problems.

By 2030 there will be 3bn people who will need to be nourished in sustainable ways.

The cost of maintaining Ethiopia's army is several fold more than feeding victims of drought in the country.

70% of Ethiopians do not have enough to eat. Half the under 5s are underweight.

CASE STUDY

Ethiopia starving again!

1969–1973, 300 000 lives taken by famine in Ethiopia. 1977–1978, 1 million flee war, drought and famine in Ethiopia. 1984 –1985, famine in Ethiopia kills 1 million. 1984, 10 000 die in food crisis.

2000 (April), the Ethiopian government has already appealed for 800 000 tonnes of food aid to feed some 8 million people threatened by starvation, but unless some rain falls soon more than a million tonnes will be needed, according to aid organisations. Ethiopia's 'disaster preparedness and prevention commission' blames two years of drought, sporadic heavy rains, frost, black-beetle and crop damage by hail for the food crisis.

But it is the war between Ethiopia and Eritrea that is really to blame this time. A UN appeal valued at some £126 million has already met with some success. The EU has pledged 430 000 tonnes of aid, the UK contributing £2.5 million of this aid. The greatest problem is through the logistics of getting food to the hungry. Aid is only being provided if the Ethiopians pledge to end the war with Eritrea. Without aid and debt relief the cycle of crisis and poverty is not expected to be broken; the population will continue to increase and inadequate farming practice will continue!

Summary

12m are at risk from starvation in the Horn of Africa.

In the world there is enough food to feed everyone adequately, but the food is unevenly distributed around the countries of the world and especially within the poorer countries. As a whole both Europe and the USA produce surpluses and

> **Possible synoptic links with:**
> - Agriculture and food supply
> - Development and human welfare
> - Population

some LEDCs actually export to the MEDCs. Some final points:

- Famine is avoidable – many countries in Africa have traditionally produced enough food to feed themselves. Changes in crops and farming methods have caused famine. A return to sustainable agriculture is a must.
- In the 'hunger' business – the USA sees the exporting of grain as a way of reducing the trade deficit and eliminating agricultural surpluses (it produces one-third more food than it needs!). However, exporting grain to poor nations can seriously harm the nations they are supposed to help.
- On aid, the poorest nations often make difficult choices when dealing with hunger and malnutrition. Accepting aid (surpluses) is no substitute for adequate food production, which would free countries from dependence on international assistance.
- On increasing food supply: (a) drainage, irrigation and terracing can increase the area of agricultural land; (b) you can increase yields through the use of fertilisers, crop spraying and by developing high-yielding varieties of seeds.

> More than 25 years after the first 'world food conference', the goal of eradicating hunger, food insecurity and malnutrition are still elusive.

However, limiting population growth will have the most immediate and dramatic effect.

11 The unstoppable loss of biodiversity

Background

> Conservation of biodiversity is divided into two groups: *Ex situ*: gene banks, tissue culture, captive breeding and *in-vitro* methods.

Biodiversity is a measure of the range of variation occurring in the natural world. This includes variation within and between species, and their ecosystems. Changing a habitat will alter the diversity of the species contained within it. A change in the number of species will gradually affect the nature of the habitat. Biodiversity is important to man for agricultural and pharmaceutical reasons.

Ecosystems and the species within them are continually evolving to cope with 'change'. Natural rates of change have been accelerated by man's activities, such as those induced by changes in land use, pollution and over-population. Accelerated change causes species extinction and ecosystems are irreversibly disrupted and the gene pool can no longer cope with change.

Amplification

> It is thought that there are cures for cancer and AIDS in the reefs.

Coral reefs are the most productive marine ecosystem. They are home to 35 000–60 000 species of plants and animals. 25% of the world's marine life lives in reefs. They are comparable to rainforests in their biodiversity. Marine communities depend on reefs for social, economic and cultural life. And they protect many coastal communities from storm waves.

> In the last four decades, mankind has destroyed 14 million ha of coral reefs. Reefs of 93 countries have been damaged by human activity.

Major threats to coral reefs include:

- sedimentation
- fishing with explosives
- human runoff
- cyanide fishing
- collection and dredging
- water pollution
- careless recreation
- global warming
- storm damage.

Marine algae is the fastest growing plant on Earth.

Jamaica's coral reef ecosystems – Negril Marine Park

CASE STUDY

The Negril coral reef has between 3% and 15% live coral. Today the dominant species are the harmful microalgae, their growth supported by nutrient 'floods' of nitrate, phosphates and ammonia (found in soap, sewage and fertiliser) exacerbate the 'local situation.' Coral needs clean, clear nutrient-free water in which to thrive! The microalgae overgrow the corals and block out sunlight.

Given the situation locals, furthermore hotel owners and the Government, have become 'Guardians of the Sea'. This group formed JCRAP (Jamaica Coral Reef Action Plan), the only known coral reef protection group in the world! Its solutions to the local problem include reassessing water quality, requiring permits to be issued for beach front building, relocating beach squatters (and their poor sanitary arrangements), providing adequate sewage disposal, planting buffer zones between land and sea to capture fertiliser runoff, watershed protection and retraction of tourism licences.

Summary

Possible synoptic links with:

- **Ecosystems**
- **Sustainability**
- **Development and human welfare**
- **Settlement**
- **Tourism and leisure**

The diversity of life within ecosystems is an irreplaceable asset to humans, serving integral roles in both social and economic structures. Much of what they do is also tangible, soil generation and productivity, water cycling and cleansing, climate control and so on! The case study above has demonstrated just one ecosystem and what is being done, but all ecosystems are under threat from man's unstoppable march forward. To enable informed decisions to be made in the future, there is a need for global co-operation in association with long-term investments to maintain monitoring of the world's ecosystems.

Sample question and model answer

1 Study the diagram below which shows waste disposal methods in the United Kingdom in 1991.

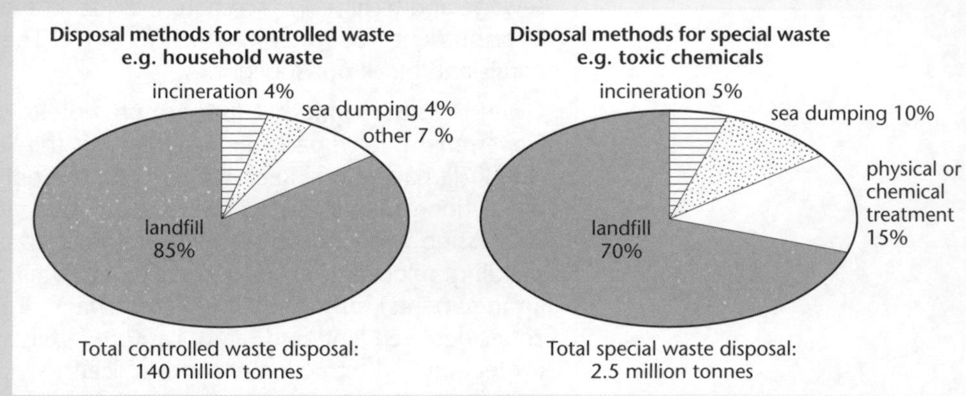

Source: adapted from DOE, The UK Environment *(HMSO) 1992*

> Landfill is an issue in the UK as we are rapidly running out of space for our waste. Newspapers are full of such references at present.

(a) What is involved in waste disposal by landfill? [4]

> Has a clear understanding of the principle of landfill

> Some impacts could have been extended

> Rehabilitation of the site?

Landfills typically involve the compaction and crushing of waste before depositing; some waste may even require treatment. Waste is often dumped in abandoned quarries and out of town sites (brownfield sites). Landfills are sometimes covered over with earth, for instance at Clacton, Essex. Edge of town and brownfield sites are used because they are cheap and accessible.

(b) Explain why landfill is widely used for waste disposal.

> Balances landfill against other methods of disposal..

> This example has more detail and as such would gain more marks.

> Might also mention transport advantages and the need for disposal sites to be nearby.

Landfill is often used because recycling of waste is so very much more expensive. Landfills are also perceived to be less hazardous than the likes of incineration (because of fumes) and sea dumping. Landfills are also relatively cheap and due to the localisation of landfill sites costs are kept down. For example in New York the city's waste is taken by truck to the periphery and deposited – this is seen as better than releasing fumes into a populous city or dumping off the coast where it would be certainly washed about in the bay.

(c) Explain why the recycling of waste material is becoming increasingly important.

> Benefits of recycling, Good.

> Too general – some detail might have developed this.

> Government policy and need to recycle to safeguard the environment for future generations.

There is a range of benefits from recycling, amongst other things reduced pollution and reduced pollutions result. Within society, especially in MEDCs, there are wider political and cultural issues that are becoming increasingly important, this being expressed in many government policies.

In relation to landfill issues loss of amenity due to smells, pollution of groundwater and methane release are now issues that won't be tolerated by today's society. Taxation and tight control ensure that to recycle saves money and is sustainable.

Assessments and Qualifications Alliance

Comment:

Not a bad attempt, but tended not to address the issues in sufficient depth. Greater use of case study and exemplar information might have helped. Awarded 10/15.

Practice examination question

How true is it to say that the less economically developed world may never be able to feed its people? [25]

Assessments and Qualifications Alliance Paper 3, 1997

Issues 12–15

12 Six million and counting – the 'Chinese census timebomb'

Background

On 12 October 1999 the UN announced that the world's population had reached 6 billion – an increase of 1 billion in just 12 years: China has the world's biggest proportion of people (though India may well overtake it by 2045).

With a population that is expected to rise from 1.3 billion to 1.8 billion over the next 50 years, China continues to embrace the one child policy introduced in September 1980 to counter population pressures created by Mao Tse-Tung's 'glorious mother' policies. The Chinese Government claims that the policy has prevented 300 million births over 20 years. But this has been at a cost: dark and evil tales of mass abortions and infanticide have abounded and it is rumoured that China's enforcers of the 'one child family' remedies are as ruthless today as they've always been.

It is estimated there are 1 000 000 unregistered in the population at least!

The reason for the renewed vigour in implementing the 20-year-old one child policy is simple. In November 2001 China will conduct its first official census for a decade. Family planning officials are petrified that the census will reveal just how they have let the one child policy slip especially in the countryside and how many extra unregistered children have been born. For many years now the family planners have taken the education and persuasion route rather than to 'punish'. But, with a looming census there has been a vicious and unprecedented backlash by family planning officials. They have to get population numbers down and quick, or risk demotion, losing bonuses, their homes and their jobs.

The population is expected to have doubled in the time the one child plan has been in place!

CASE STUDY

Baby traffic beats 'one child' policy

This traffic in which poor families from the hinterland sell surplus infants to the richer coastal provinces, is taken for granted.

'Girls are less popular.' They only fetch 1500 renminbi (£120). A good healthy boy is worth four times as much. The official *People's Daily* has reported that almost 10 000 women and children are abducted and sold each year in Sichuan alone. Although no nation-wide figures are available, the practice is known to be prevalent in such other poor areas as central Hubei, eastern Anhui, southeastern Jiangxi and southern Guizhou.

A boy is more useful because of his labour power. As he grows up he can earn more in the fields, and even more if he becomes a migrant worker in the city. The traffic illustrates one of many loopholes in the 'one-child-family' policy. Police have now cracked down on baby selling but strong economic incentives may make it hard to eradicate.

For countryside dwellers this is 'easy cash'. 'Have a fat tummy every year; after three you'll be able to cheer'! Alarmed by the problem, Beijing has launched a campaign against selling women, which it has classified as one of the 'six-evils', along with prostitution, pornography, gambling, drug trafficking and what it calls feudal suspicion (consulting fortune tellers and the like). During a 15-day crackdown in the coastal province of Shan-dong last June, city and court officials were called into the governor's office and told that they would be held responsible if they failed to take action. During the 15 days, according to the English-language newspaper *China Daily*, 170 women and children were rescued, 859 middlemen arrested and 82 gangs broken up.

Summary

2000 is the year of the dragon. In dragon years it is lucky to have a child.

'As long as the FPO have quotas to meet, they will use whatever means at their disposal to stop women having children.' And there is no doubt they are effective at it. Let's hope the results of the censors 'don't give the FPO the excuse to get even better'!

Possible synoptic links with:

- Population
- Migration
- Development and human welfare

CASE STUDY

'Do you realise it is illegal for this baby to live?'

For many it is not the one-child policy that's the problem but how it is implemented: 'orphanages are swelling with babies left to die by their parents. 90% are girls, and every year 750 000 go unaccounted for'. Though the care offered in these orphanages has improved they are still dubbed the 'dying rooms'.

Those that fall pregnant can be treated with absolute disdain and inhumanity:

- in China iodine or formaldehyde is injected into baby skulls as they 'crown', they die instantly, and in China this is classified as legal abortion!
- new born babies are drowned or beaten to death!
- in some cases, pregnant women who have violated the one child policy are held in detention centres by family planning officials whilst attempts (with injections of saline into the foetus) are made to kill the unborn child.

Those who fall foul of the FPO are ostracised, beaten and fined.

13 'Trade not aid'

Background

Countries can improve their standard of living by selling more goods abroad (exports), this gives them money to buy goods from other countries (imports). Many countries make it difficult for others to export their goods to them by putting a quota, or limit, on the amount of goods they import and having tariffs, or taxes, which are added to the price of imported goods.

The EU is the worlds biggest trading bloc.

The EU has the third biggest economy in the world and the world's biggest aid programme. Since 1975 the EU has linked trade and aid together under the Lomé Convention, the aim of which is to develop in countries in Africa, the Caribbean and the Pacific (ACPs) 'a partnership agreement based on mutual independence'. Put simply, Lomé guarantees African, Caribbean and Pacific countries special access to European markets with no reciprocal access for European exporters.

Lomé has two key components: Aid and Trade. The EU's aid policy is known to be successful because of its focus on development rather than strategic and commercial interests. On the trade angle it has been successful too: with no tariffs to pay, Africa, the Caribbean and the Pacific members have enjoyed a competitive advantage over say Japan and the USA.

Given the 'life support' role that Lomé has provided, ACP countries ask why after 25 years is it to be removed? Principally, it is because of poor infrastructure within ACPs it means there's a lack of development initiative and little entrepreneurial input, ACPs are poor countries. By focusing dependence on the EU, ACPs have not sought out other markets. Globalisation of industry means that the advantages of free access provided by the EU have diminished. Many agree that locking the ACPs into Lomé has been unfair.

WTO consider Lomé to be unfair.

A stay of a further 2 years was agreed beyond 2000 for Lomé late in 1999. After this period Lomé will be replaced with regional free-trade areas, this will reform reciprocity to importing and exporting, and tariff barriers will have to be removed.

CASE STUDY

Is the Lomé pact crucial to Pacific Island ACP Nations' Survival?

Nothing less than the economic survival of island nations in the South Pacific may be at stake as the European Union continues to consider the revision of a special trade and aid pact with African, Caribbean and Pacific (ACP) states.

The lifeline of some of the Pacific island economies, such as the preferential access the canned tuna industry gets to the EU market, is among the key areas up for renegotiation. Tuna is the biggest export earner for most of these countries, this industry is crucial to the long-term survival of most island nation economies.

Vanuatu's fledgling beef exports and tree crops are other areas which will be badly affected if the Lomé Convention is not renegotiated. So will Fiji's sugar industry.

The eight signatories to the Lomé Convention from the South Pacific are Fiji, Kiribati, Papua New Guinea, Solomon Islands, Tonga, Tuvalu, Vanuatu and Western Samoa.

Since the Lomé Convention was negotiated in 1975, the eight Pacific island countries have received EU aid worth $1675 billion.

The biggest, Papua New Guinea, with a population of $4.5 million, has received $725 million while Tuvalu, with 9000 people, has benefited from $12 million in aid and trade benefits.

The EU's ambassador to the Pacific, Germany's Gerd Jarchow, has hinted that the EU may not be that committed to the Pacific in the coming negotiation phase.

'Lomé has to be changed because the world has changed so much'. 'You can't treat the Pacific the same way as Africa or the Caribbean countries. The interests of the Pacific is more to the Pacific Rim countries than to Europe.'

Further, now that central European nations are about to join the body they will have no interest in paying for the Pacific. They want to pay for themselves and this could have an impact.'

A typical case in point is Fiji's sugar industry, where Europe pays, under the Sugar Protocol Agreement, a guaranteed price which is sometimes three or four times above the world market price.

This agreement alone is worth some $141 million a year to Fiji. Sugar exports account for 12% of Fiji's Gross Domestic Product (GDP) and 58% of its agricultural production. In a country of 750 000 people, the sugar industry provides an estimated 105 000 jobs.

Though Fiji's sugar agreement with the EU technically goes beyond the 2000 deadline for the Lomé Convention, it is likely the agreement will be removed as Lome ends.

Working on this premise, Fiji has planned its economic strategies several years ahead. Whether this confidence is misplaced remains to be seen.

Other Pacific Nation members are hoping for the EU to throw out special lifelines for their industries when the Lomé Convention is renegotiated!

Possible synoptic links with:

- **Development**
- **Agriculture**
- **Industry**

Summary

The established end of Lomé will undoubtedly damage developing countries, unemployment will increase and the standard of living will drop, as ACPs have to face up to the fiercely competitive world economic market. ACPs will undoubtedly be further marginalised occasionally and socially; appeasing WTO may spell the death knell for many of the ACPs!

14 Europe 1945–2000 and Monetary Union

Background implications

The EU's GDP in 1999 = $8 458 billion.
The USA's = $9 190 billion.
Japan's = $4 380 billion.

The European Union has its origins in the European Coal and Steel Community, a union of six nations, established in 1951. In the year 2000 it is a union of 15 nations; this could well have doubled by 2020. The Euro and Schengen accords (open borders) have undoubtedly brought nations closer in Europe, with increasingly developed local and regional power relationships between the EU's diverse membership assured. Involvement in Europe brings huge trade benefits; despite the globalisation of business, EU members do most of their trade within Europe. However, political ambivalence has weighed heavy around the EU's neck since its inception, in terms of foreign and security policy.

CASE STUDY

The single market and the Euro

EU members signed the Maastricht Treaty in 1992. Amongst the many things established, was a single economic and monetary market and currency by 2000–2002, the aim being to bring price stability to the ever-enlarging European Union.

In January 1999, 11 of the 15 members of the EU joined the EMU (European Monetary Union). The UK, Sweden and Denmark opted to stay out (Greece didn't meet convergence criteria – but will join the EMU in 2001). If the EU enjoys a major boost and trade and GDP increases, Sweden, the UK and Denmark (whose peoples at present are all broadly opposed to the Euro) may well adopt it.

In the first year and a half of trading the Euro has proved to be rather weak against the dollar, losing 20% plus of its original value since it was introduced.

In 2002 it completely replaces the natural currency of those under EMU at present. All financial business will then be subject to the financial disciplines of the ECB (European Central Bank).

The EMU/ECB and Euro will have massive economic and political clout, and the UK, Sweden and Denmark will surely think differently about their membership of the EMU. Already some big businesses are trading in Euros to ensure continued business footholds and stakes in the rich European Markets.

Summary

Possible synoptic links with:

- Agriculture
- Population
- Development

At least 12 other European States have shown great interest in joining the EU. We live in a rapidly changing Europe and one that will increasingly shape our lives.

Three important European groups will in particular increasingly influence our lives through the next century.

- **WEU** the Western European Union was first formed as a defence grouping in 1948, and it has been revived as Europe's Security Force. It consists of a 60 000-strong 'rapid reaction' force, of European multi-national forces. It's thought it will replace NATO's role in Europe in time.
- **COREPER** (the Committee of Permanent Representatives) is Europe's top decision making body made up of national ambassadors who work to bring to

fruition the dictates of the European Council, the heads of government and foreign ministers of member states.

- **EP** (European Parliament) has no law-making ability, but does control the EU budget.

15 Transport congestion

Background

It is estimated congestion costs the UK £10 million/year.

Know the terminology!

Transport has become a key political battleground and one that seems unlikely to go quietly away, given the high taxation on road use and seeming lack of investment. The congestion on today's roads has many causes. Primarily two decades of planning mainly concerned with predicting and providing based on estimated future car usage, has resulted in massive road building and consequently encouraged more and more cars to appear, e.g. M25 (intended for 80 000 vehicles now has over 200 000 vehicles a day).

Bangkok, Tokyo and Mexico City are worse.

Secondly, deregulation and privatisation of public transport services, allegedly to regulate and ensure continuity to encourage competition and innovative transport forms and therefore increase passenger demand, has had completely the reverse effect, forcing the travelling public back into their cars! The present thrust of Transport Policy thus comes at a time when in the UK there are too many cars on the road, resulting in gridlock and increased air pollution, and when public transport has undergone serious neglect and decline. Also as Britain is getting richer and we expect to travel where we want when we want and complain when congestion means we cannot!

Why are we acting now?

In 1979 there were 70 cars/mile of road, by 2000 there were 105 cars/mile of road.

Principally because world traffic is expected to grow by 30% in the next 20 years. Policy is also aimed at halting the increasing lack of use of public transport: today, in excess of 80% of UK motorists never use a bus. Further, over the last two decades or so, less than 1% of GDP has been spent on transport. But also there is concern that the economy will suffer and votes in future elections will be lost.

What is happening? And how will the money be used?

Ensure you know some examples of such schemes.

The immediate change is that transport spending is increasing from £5 billion to £9 billion in what is a 10-year plan which will, it is promised, 'transform the continued transport network'. The success of solving the congestion problem will obviously hinge on how effectively and where money is spent on transport. The extensive park and ride schemes and light transport schemes, in addition to the obvious input into more public modes of transport, are areas that one would expect to be at the core of policy. Additionally it's expected that schemes will be established that rapidly reduce the numbers of cars on the road. By increasing the real cost of car usage, through tolls, taxes and cost increases on parking etc., improving the motorways with closures of motorways to those making short journeys, through variable speed limits and the extension of crawler lanes. Alternative methods of transport will also be highly subsidised and encouraged, e.g. bicycle lanes. One of the first steps at identifying priorities has been the requirement on local administration areas to draw up Local Transport Plans, money will undoubtedly be allocated based on such plans.

The best light rail system in the world?

Metrolink opened in Manchester in 1992 at a cost of around £152 million. A fleet of 26 trams operates over the 31 km network. Metrolink has now reached 13.9 mn passengers a year and 65% of Metrolink passengers have a car that they could have used instead of Metrolink. Between 14% and 50% of car trips to destinations served by Metrolink have been switched to Metrolink.

Why is it so successful?

The system is simple and easily understood. Quick journeys and city-centre track gives good access to the main attractions and work places. The service is frequent and reliable. The system is safe to travel on.

Extensions through Salford Quays have been built running along Eccles New Road, and will serve a significant residential population. Four further extensions to Oldham and Rochdale (cost £137 million), Manchester Airport, Trafford Centre and East Didsbury, were approved by the Government.

In a very short time Metrolink has established itself as a very successful transport system, tempting people out of their cars in a deregulated bus environment and without subsidy. It is part of any integrated transport strategy for Manchester. It is a clear indication that investment in good quality public transport works.

CASE STUDY

Possible synoptic links with:

- Settlement
- Ecosystems

Summary

In the short term the measures that are being undertaken will mean that the jams go. But beyond this period, transport investment invariably means that the jams will get worse; but at a different location and in a different town to that hit now! Let's just hope that the £700 million pledged to the Channel Tunnel link (in a £5.6 billion project) does not result in nothing being on offer to other priority projects!

£700 million would be enough to build bus lanes in every city in the UK.

Sample question and model answer

1

Study the diagram, which shows the direction of change of international trade of the member states of the European Union.

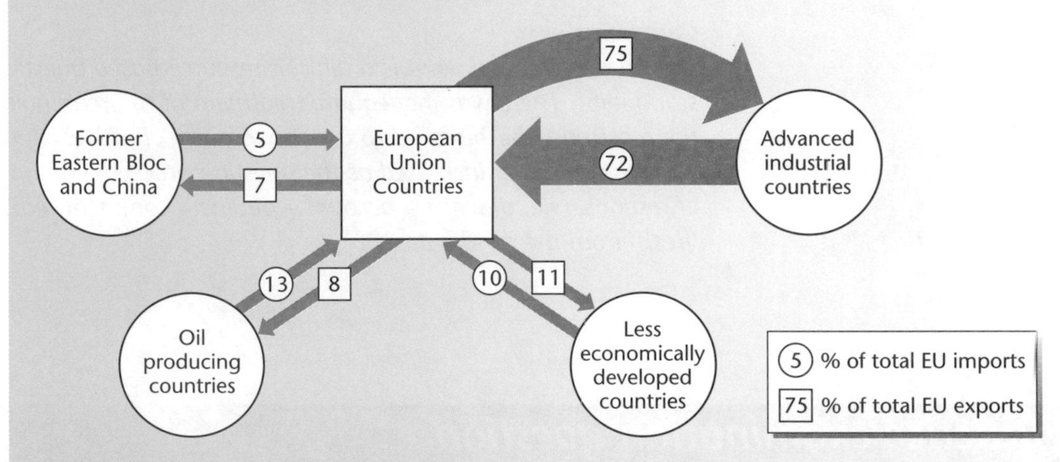

Source: adapted from M.Witherick et al., Environment and People *(Stanley Thornes) 1995*

(a) Describe the pattern of trade shown in the diagram. [4]

Makes correct observation in first line.

Recognised the imbalance between LEDC's and MEDC's.

Trade from Eastern Bloc, OPEC countries and LEDCs is far less than that between Advanced Industrial Countries and EUC countries. For instance exports to LEDCs are 11% whereas to Advanced industrial countries they are 75%. Imports are 10% from LEDCs and 72% from Advanced industrial countries.

(b) Account for the relative importance of the trade with advanced industrial countries compared with less economically developed countries. [4]

This answer rather lost its way. There was no reference with regard to lack of trade with the eastern bloc. Europe as a trading bloc – not covered. What do LEDCs make that is needed in great quantity?

Trade with LEDCs is important as very often LEDCs possess large amounts of raw materials that can be acquired cheaper than from anywhere else. However, advanced industrial countries, e.g. USA, Japan, have much better technology and money, their investment and trade is good for Europe e.g. UK-NE and Nissan.

(c) Outline the benefits to a less economically developed country of increasing its trade with EU countries. [5]

Simple comment to the relationship between levels of development and what's produced.

Changing dependency.

Multiplier effect.

Idea of inward investment. A better answer but still lacks specifics.

LEDCs have in general many natural resources but by comparison their manufacturing industry and their ability to create high quality, high technology goods is very poor. By trading with EU countries (UK, Germany, France) they can obtain higher technology goods and machinery to aid them with their development and raise their standards of living. It is also important as foreign investment (often by private companies and multi-nationals, e.g. Ford) provides money and labour for the country. Paradoxically it has the effect often of draining money out of the country to multi-national HQs.

(d) Comment on the view that economic development always leads to an increase in human welfare. [7]

Use 'indicators' to elaborate the process of development.

Material consumption increases.

Economic development has its benefits; mainly increased amounts of money leading to better education, healthcare, amenities, luxuries, and better public services. This has been especially true in the development of western countries. On the other hand, development by multi-nationals in Africa has often been in their own self-interest rather than the welfare of the indigenous population, e.g. oil companies such as BP have interests

Sample question and model answer (continued)

Drawbacks of accelerated development elaborated upon here.

in Nigeria, but although the economy has developed much of the money is simply drained back to Europe and some of Nigeria's poorest people live in the oil-producing area of the coast.

Comments

This student tackled what is a difficult resource-based question from a rather simplistic standpoint. There was inadequate treatment of the terminology of the subject and for the most part she/he failed to use the resource effectively enough. The required level of response is overtly increased as the question proceeds. Command words 'up' the level of response: i.e. describe – account – outline – comment, each demands something 'fresh' from the candidate.

Practice examination question

Spatial variations in the physical environment create both economic problems and economic opportunities. Discuss with reference to regions in the EU. [25]

Source: OCR Geog A – Specimen Question

Practice examination answers

Chapter 1 Cold environments

1 You must look at high altitude and high latitudes. An understanding of how this topic links with the rest of the geography core will be sought.
There are three areas to this question.

- Describe the constraints imposed by the environment; the difficulties associated with limited economic activity and low densities of population need to be investigated. Resource exploitation and improvements in technology related to this environment must be investigated. The strategic, research and tourist potential also needs to be covered.
- The success of strategies in the periglacial realm needs to be assessed, i.e. has the life of indigenous tribes become any easier? Has technology made life easier? Has accessibility been improved? Are resources being sensibly exploited?
- What are the costs of the development in these areas? i.e. cultural, environmental, impact on the permafrost itself, waste and litter etc… The fragile element of this environment needs to be developed.

The best answer here will be balanced, but also extremely evaluative.

2 (a) You must demonstrate an understanding of how glaciers advance and retreat and that spatial variations do occur. See the relevant text on page 23.
(b) You must emphasise that every landform is distinctive and that all landforms change over time. The examples that you cover must demonstrate a good range of erosional and depositional landforms. Diagrams, maps and plenty of good located examples are very necessary here. Pages 24–29 in this book will help you.

Plenty of appropriate terminology will be very convincing in this question.

Chapter 2 Tropical environments

1 You should use examples that demonstrate outcomes of the interactions of physical and human processes, they should show a clear understanding of sustainability, and look at long-term implications.
Answers on the whole are likely to identify:

- Chopping trees = a break in the nutrient cycle because stores are removed.
- Chopping trees = disrupts other stores and transfers.
- Chopping trees = you reduce soil fertility therefore the numbers of species of trees and all other plants are reduced.
- By chopping down trees = removal of habitat.
- Tree removal = increases in run-off and flash flooding.
- Modification of climate.

2 You need to offer a range of benefits, recognising that benefits can be both economic and environmental.
Examples include:
Subsistence needs – fuel wood, building materials, fodder, nuts, medicine, dyes
Environmental uses – limits soil erosion, shade made available, wind breaks
Industrial uses – veneers, gums, oils, gene store, medicine
You should show a clear understanding of sustainability and develop your answer beyond the obvious benefits above. You might well refer to the consequences of using the forest resources beyond the locality.
Answers that dwell on a single resource or just look at the effects of deforestation would not score highly!

3 A range of answers is possible here. But in summary, your key points must emphasise that LEDCs and MEDCs have global as well as local/national responsibilities. You would need to show a keen awareness of a range of development issues.
You should be aware that there are differences between LEDCs and MEDCs. The fate of forest areas is for instance very different in the eyes of those that live in LEDCs from those that have all the trappings of a Western life style. You must use detail to support your case and must stick to a clear introduction, development and summary structure.

Chapter 3 Arid and semi-arid environments

1 (i) Look at the diagram on page 61. The main features are wadis and gullies, alluvial fans, pediments, bajada and playa. There needs to be some appropriate scale to the diagram. Descriptive and accurate annotation will also be important here. The best answers will undoubtedly take descriptive comments beyond the obvious.
(ii) The past will be linked to the present climate. The fact that running water has influenced this landscape will be fully understood; in particular, water's link to pedimentation, duricrust formation, desert varnishes and the dissection of plateaus will be important. An awareness of the theoretical composite nature of landscape will be understood.

2 (a)
- High pressure all year
- air is descending from height
- descending air warms as it falls
- the air above the area is essentially stable
- prevailing easterly winds blow inland
- cold currents to the west of the continent.

(b) Most recent event was the last ice age, pressure belts and wind systems move south as ice advances. There would have been low pressure/depressions and westerly winds all bringing rainfall.

3 Causes

It's arid because:

- of continentality
- circulations in Hadley Cells
- coastal cold currents.

Rainfall, when it comes:

- relates to relief
- relates to convectional activity
- is sporadic and short lived.

There is a latitudinal effect, and there tends to be more rain on western coasts.

Consequences

- Impact on denudation – weathering is either aeolian or based on sporadic fluvial activity.
- You could cover the variety of landforms found and any associated ephemeral vegetation.
- Impact on traditional society. Pastoralism and nomadism could be covered. Talk about water management and irrigation (salinisation) and the search for new water supplies.
- Settlement prospects could be covered.

Chapter 4 Migration: current issues

1 (a) (i) A town or region, a holiday resort or retirement area, could be coastal or rural [1] (2nd and 3rd mark for some data from population pyramid)

(ii) declining area, e.g. through mine closures or a stagnant rural area [1] (2nd and 3rd mark for some data from population pyramid).

(b) Benefits include:

- employment
- more nursing homes
- increasing spending power (of elderly)
- town is growing, favouring the construction centre.

(c) Possible consequences include:

- racial tensions
- cultural tensions
- social segregation in towns
- lower wages paid
- unemployment
- population structure unbalances
- cost of education/social security/etc.

Chapter 5 Development issues

1 A

(a) (i) Core – A concentration of economic power, wealth, innovation and productivity. Will dominate in political, economic and social aspects.

Periphery – A less developed or undeveloped area which may serve the core area with resources (by exploitation) and by people (by migration).

(ii) Look at distribution and development. In the core areas the distribution is uneven, though most is concentrated in the Northern Hemisphere. There is a continuous band through USA, Europe, Japan, etc. Australia is anomalous in the Southern Hemisphere. In the semi-periphery areas the pattern is more fragmented with some significant concentrations in E. Europe, Latin America, South East Asia, etc. Saudi Arabia and South Africa are anomalies.

(iii) Areas where GDP and the use of energy have risen but still large percentages employed on the land and very uneven distribution of wealth. Using models and percentages employed or birth rates and death rates as indicators of socio-economic progress.

(iv) Areas which are 'taking off'. There may be a relation to former colonial occupation. Proximity to major core area. Your answer must balance physical, economic, historical and political comment.

(b) Physical factors can be seen as both advantages and disadvantages. In terms of resources, accessibility, proximity to core areas and the potential for both subsistence and commercial existence has to be covered.

B

There are two parts to this question, 'why and how'. You need to outline that government policies can be both direct and indirect. You must also cover the topic at a local and national scale. So the best responses will use details of policies to explain the process of regional development, and will appreciate the way in which processes operate at a range of scales. The word 'region' could mean any area at any scale. Specific, particular and actual countries are needed in this answer.

2 (a) The very best answers will make a number of suggestions relating export earnings to comparative advantage and the possibility of subsequent diversification.

(b) To reach the highest level you will make a clear and fluent distinction between trade and aid. Your answer will also display an understanding of the changing nature of world trade, and that includes trade blocs. Some good contemporary exemplars need to be used too, to make your answer completely convincing.

Chapter 6 Leisure and tourism

1 (a) (i) The numbers curve flattens in the late 1970s, early 1980s and income actually falls. (1) for the date, (1) for what happens.
(ii) 'How' is easy. Everything went up! (1)
For 'why': changes in transport technology and changes in society, too, from more income to higher expectations (4 marks).
(b) The extract shows that there are problems caused by tourism in Kenya in so far as the chief attractions, the wildlife, cause damage to farms' crops and sometimes to farmers, too. Farmers do not see much of the money generated. Wildlife tourism is the country's chief foreign exchange earner and, so, absolutely crucial to the government for all its spending needs, not just for services for farmers. To encourage tourism is necessary for government income, but to have the tourists you have to have the wildlife that causes such problems to the poor farmers.
(c) (i) What is the country's or region's policy? (1) for identification, the other (3) for the description.
(ii) Tourism is the world's biggest industry and it is vital to many countries' employment prospects. There are many negative consequences of it. What this question required is to take their case-study country's policy and to assess the policy's role in gaining most overall benefits for tourism.

Chapter 7 Agriculture and food supply

1 You must attempt to show the connections between the various parts of the specification: ecosystems, population, rural settlements, agriculture and the EU. You should describe the impact of the agro-ecosystems on soils, wildlife, water resources, rural landscapes. Show how farm policies are linked to environmental damage. Are environmental 'costs' acceptable? You must be synoptic in your approach and reference real places. Synthesis and evaluation will be obvious in your answer.

Chapter 8 Synoptic assessment

Issues 1–3

(a) Your diagram will show:
- knowledge of major gases involved, and the processes occurring
- an understanding of links
- your annotations will aid the answers.

(b) Your evidence should link global warming to 'enhanced greenhouse' effects. That it can be linked to volcanic erupting fires and sunspots and to human activities (fossil fuels, burning, landfill, methane release).
- Your answers will show a sound link between greenhouse effects and global warming, assess conflicting evidence and demonstrate something of the scientific controversy surrounding the topic.

(c) Choosing: rising sea level, droughts, increased rainfall/storms and loss of biodiversity; you will assess the consequences of global warming and the need for scientists to study the phenomenon.

Issues 4–7

1
- Green Revolution comments are okay, but its success is patchy, e.g. Africa's experience is very different to India's, Africa's food production continues to drop!
- Climate comments – to do with drought, flood and cyclones.
- Soil and ecosystem comments – to do with erosion, pests and salinisation.
- Development comments – cash-crop dependence, poverty and literacy limiting knowledge and therefore development and land-ownership changes.
- Agricultural – methods are both outdated and high-tech, this creates massive problems.
- Population – pressure of growth. Concentration of growth in cities therefore an absence of labour on the land.
- Hydrological comments – to do with lack of/or poor availability of water.

Perhaps exemplified by ENSO, drought, agricultural methodology and population pressure studies. Global, regional and local exemplars are fine here.

2 Basic comments, to do with food production increases through extensive and intensive methods together with examples of consequential environmental impacts are needed here. Further, the question requires you to understand how agriculture is meeting the demands of increased population and how it affects the environment. Relevant is:
- The adoption of more advanced techniques, through greater capital and aid input, can lead to a loss of biodiversity (deforestation, over-fishing and loss of wilderness areas); well-building to increase water quality and quantity, leads to groundwater depletion and salt ingress.
- Remember environmental impacts can be good too, lessons are learnt from previous mismanagement! As population growth slows, the environment is not so heavily exploited; there may be an increase in literacy, this helps the environment too!

You need to exemplify improvement methods (green revolution, technological methods) and environmental problems created (biodiversity loss: in forests and seas? etc.).

Chapter 8 Synoptic issues (continued)

Issues 8–11

A true A2 question in many ways. The theme is explicit and extremely evaluative in nature. The best students will offer a balanced answer which is reasoned and considered. There are two threads to the answer.

Causes of Famine

Wars/political problems
Desertification
Climate Change
Land ownership problems
Poor farming methods

- This suggests the Developing world is failing – with underlying and persistent problems of malnutrition, health and poverty.
- Africa is worse affected!
- Rural urban drift results.

Signs of hope

- SE Asia famine has been eliminated.
- The green revolution has worked.
- Birth control is working in many LEDCs
- Long-term aid

Issues 12–15

The best A2 answers will connect the different parts of the Specification you are studying, for instance, the physical environment, rural settlement, agriculture, industry and tourism and so on. A discourse of a range of physical environmental factors, e.g. relief, hydrology, soils, geology and climate in the Mediterranean, high latitudes, etc. will be looked for. Isolation, remoteness, mineral and energy resources, climate and soil resources will be discussed. Your answer should be balanced in terms of problems and opportunities. Your synoptic answer will have plenty of examples and all ideas will be well synthesised.

Index